BASKET WEAVING

Toon Brekelmans

BASKET WEAVING
Toon Brekelmans

Translated by David Wright

ISBN: 978-1-914934-98-8

Basket Weaving (Korfvlechten) 1st Edition published in Dutch, 1979; 2nd Edition published 1985; both by Cantecleer. 3rd Edition published 2002 by VBBN.

1st English edition 2025 © Northern Bee Books, Scout Bottom Farm, Mytholmroyd, Hebden Bridge HX7 5JS (UK). 01422 882751.

While every effort has been made to contact the original copyright holders this has been unsuccessful. The publishers would welcome information from holders of these rights.

English translation 2025 © David Wright.

Front cover image © David Wright: a skep decorated for a memorial service for Phil Moss (1937-2023), a much-loved beekeeper. Rear cover image © Ben Sunderland: a skep with Mull in the background.

Book design by www.SiPat.co.uk.
Typeset in Baskerville and Minion Pro.

All rights reserved. No part of this publication may be reproduced, stored or transmitted in any form or by any means electronically or mechanically, by photocopying, recording, scanning or otherwise, without the permission of the copyright owners.

BASKET WEAVING

The spiral braiding technique for making bowls, bee-skeps and baskets

Toon Brekelmans

Translated by David Wright

CONTENTS

Introduction 5
Acknowledgements 6
Biographical Note on Toon Brekelmans. 7
Preface. 11
1   Old basketwork. 13
2   Materials and tools. 19
3   The technique of basket weaving 31
4   Bee-skeps 55
5   Handles. 65
6   Embellishments. 77
7   Bibliography. 85
8   Colophon 86
9   Additional image credits. 87
10 Resources 88

INTRODUCTION

Toon Brekelman's book on Basket weaving, (*Korfvlechten*), was first published, by Cantecleer, in 1979. A second edition followed in 1985 and a third edition, published by VBBN, in 2002. A small section of the book was translated into English by Marie-Jose Smit and, with some of the illustrations and an introduction by David Chubb, was published by Bee Books New and Old in 2000, with the title of Skep-Making.

Father Brekelmans, a Dutch Catholic priest, died in 2015, (for more information on his life, see the Biographical note).

In 2024, before starting on this translation, I made efforts to investigate copyright by attempting to contact Cantecleer, the original publishers, and to see if I could find any relatives of Toon Brekelmans. I am very grateful for help in this from Marga Canters of NBV, the Dutch Beekeepers Association, (the successor to VBBN), who put me in touch with Kees van Gorkom, an old friend of Brekelmans. I found that Cantecleer no longer exists and that no relatives could be traced. However, Kees van Gorkom encouraged us to go ahead, saying that he believed Toon Brekelmans would have had no problem with a translation. This translation is dedicated to the memory of Toon Brekelmans, together with grateful thanks to all those who contributed drawings and photographs to the three Dutch editions of *Korfvlechten*.

ACKNOWLEDGEMENTS

Maureen Campbell of Edinburgh Beekeepers first drew my attention to the possibility of making skeps and I subsequently attended skep-making courses with Bryce Reynard, with Chris Park and with Nick Mengham. The late Martin Buckle was an inspiration, both as a skep-maker and as the producer of a website, sadly no longer available, which provided detailed information on many aspects of skep-making. Marga Canters and Kees van Gorkom gave me encouragement to translate. Pierre Sanders very kindly obtained a copy of *Korfvlechten* in Dutch for me and checked the translated text. Bryce Reynard reviewed the text and made many helpful comments. Finally, I am most grateful to Jeremy Burbidge of NBB for his enthusiasm and regular encouragement and to Simon Paterson for his help in producing the book. My wife, Bron, has been a constant and much appreciated support.

BIOGRAPHICAL NOTE

Toon (Antonius Johannes) Brekelmans, the author of *Korfvlechten,* was born on 16 December 1930 in Schoorstraat, in the village of Udenhout in Brabant, Netherlands, where his father kept bees on the family farm. He was ordained as a Catholic priest in 1958, (see Fig 1), and then studied in Rome, receiving a doctorate in church history in 1965. Returning to the Netherlands, he taught church history at the Joint Institute for Theology in Tilburg, and at the Bovendonk seminary in Hoeven. He spent many years living in a monastery in Goirle, where he kept bees. His father had kept bees in skeps but he learned skep-making in 1965–66 from Gart von Gorkom from De Moer and worked on the technique over the years. He also ran skep making courses, (see Fig 2) which shows him on the left.

The first edition of *Korfvlechten* appeared in 1979, with a second edition in 1983. A revised edition was issued in 2002 and it is this version which has been translated. Apart from his work in church history and writing *Korfvlerchten,* Brekelmans made regular contributions to the Dutch Beekeeping Association journal.

Toon Brekelmans moved to the Monastery of the Missionaries of the Holy Family in Teteringen, when the Monastery at Goirle closed, and he died there on 11 September 2015.

Fig 1. Father Brekelmans (second from right) celebrating Mass in 1958. *https://wikimiddenbrabant.nl/Pater_Toon_Brekelmans#/media/File:7.7_Eerste_mis_Toon_Brekelmans_3-8-1958.jpg*

FURTHER INFORMATION

1) Toon Brekelmans, Professor and Beekeeper. An interview by Ton Thissen. *https://library.wur.nl/ojs/index.php/bijenhouden/article/view/12151/11654*

2) Learned Scientist [and] beehive authority, by Paul Spapens (from an article which first appeared in the Brabant Dagblad, 21 Jan 2003). *https://www.bijenhouden.nl/phpbb/viewtopic.php?t=13878&start=10*

3) Keeping bees, by Kees van Kempen, Frank Scheffers and Cees de Werdt *https://schoorudenhout.nl/wp-content/uploads/2020/07/Familieboek-2-Pijnenburg-Smetsers.docx*

Fig 2. Skep-making workshop in Ulvenhout, North Brabant, The Netherlands. The author is on the left.

Fig 3. The Beekeepers by Pieter Bruegel the Elder 1565.

PREFACE

The first edition of this book on basket weaving, [which was] the only one in the Dutch language, appeared more than twenty years ago and has been sold out for a long time.

The interest in this old handicraft and the continuing demand for a suitable manual made me decide to republish it with some additions and improvements.

Although techniques of basket weaving may differ from place to place, I offer you a method which I have found to work and to be effective.

If we want to distinguish this ancient craft from the weaving of baskets with wicker, we can call it spiral weaving, because in our weaving technique one bundle of straw is continuously spiralled on top of another and tied to it with a weaving band.

When describing this technique, I attach great importance to insight. Reasons are repeatedly given as to why one should weave this or that way. This is because it is only after fully understanding the techniques that a weaver is able to use their imagination to create new designs and shapes. Working on this stitching technique will bring joy to the hobbyist who is prepared to invest the time necessary. Then the weaver will take delight in a successful piece of work and also in developing mastery of the material. Josef Wetyns, the founder of the Flemish Outdoor Museum in Bokrijk, states enthusiastically. "In particular, the splendour of the regularity of the stitches in newly made baskets has a simple, surprising charm. An additional fascination, among other things, is the fact that the strength of the hand is expressed in the [neat coils] of the straw. When we see this, we feel that the human hand has overcome the resistance of the material. These are examples of a real handicraft, simple and clean"

Finally, I would like to thank the beekeeper-skep-makers who taught me, as well as the Association for the Promotion of Beekeeping in the Netherlands, and Het Biejenhuis in Wageningen, who have made this re-issue possible.

Toon Brekelmans
Goirle 2002

Fig 4. The Lovesick Maiden by Jan Steen (1626–1679). In the bottom left-hand corner is a small woven basket.

1. OLD BASKETWORK

Let's take a look back at our heritage, because basket weaving is an ancient technique. We can follow traces of spiral braiding or stitching as far back as prehistoric times. So-called pedestal and bell beakers have been found in the Netherlands that have survived from the Neolithic period (4500–1600 BC).

On the outside they have more-or-less the same pattern as the spiral stitching that we use today (see Fig 5). The potters found this pattern so decorative that they engraved it on their cups and urns.

Fig 5. Neolithic bell beaker (National Museum of Antiquities, Leiden).

Furthermore, there must have been another connection between pottery making and basket weaving. The first potters did not use a wheel, instead they made long rolls of clay or placed clay on top of more clay in a spiral. Then the mould was more-or-less smoothed out and baked. According to research these facts indicate that the earliest inhabitants of the Netherlands, before they understood the art of making pots, had the skill of weaving baskets.

Because the material didn't survive, no prehistoric and Middle Ages baskets have been preserved. Important sources for finding old basketwork are prints and paintings. Beehives are often depicted in medieval art. We have been unable to find any other spirally bound material from that time, although it must certainly have existed. We have to wait until the sixteenth and seventeenth centuries to find such material depicted. A few examples include the following. In the painting *The Fall of the Rebel Angels* by Peter Bruegel the Elder (1528–1569) we can see an oval basket worn under the arm. In Bruegel's print *The Meagre Kitchen* we see a cradle, on which a woman is sitting, feeding a child. The cradle is a kind of reclining mat with one high raised edge that serves as a backrest.

[The] painting[s] by Jan Steen (1626–1679), [The Lovesick Maiden], (Fig 4), shows a high cylindrical basket with a lid.

In his book Brabants Recht of 1628, Jan Baptist Chrystin makes an inventory of the objects that there might be on the farm and in the house, in connection with the law of inheritance. This inventory includes a straw measuring basket [a bushel] and a straw corn basket. The cylindrical measuring basket had a slat in the top edge and was calibrated by the calibration masters of the city. It was a so-called dry measure, and was used to measure grain, seed and meal. A bow was needed to level the contents evenly with the edge.

With the straw corn baskets, we come to the subject of stock baskets, many copies of which have been preserved and which will be discussed shortly.

The most important sites for [preserving] old basket-work are folklore museums and collections of antiquities. Such basket-work was generally made in the nineteenth and early twentieth centuries and comes from farms. In the Netherlands Open-Air Museum (National Museum of Folklore) in Arnhem and the Flemish Open-Air Museum in Bokrijk, there are many storage baskets (see Fig 6).

1. OLD BASKETWORK

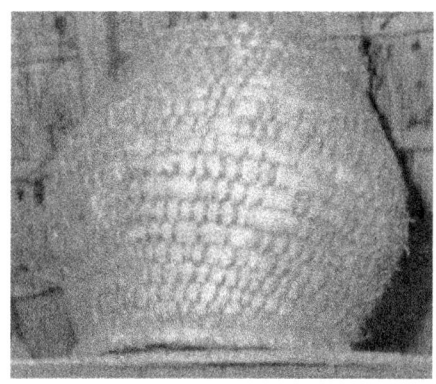

Fig 6. Storage basket. The Dutch Open-Air Museum, Arnhem.

[Such baskets] are of various sizes, some cylindrical, others tapering. Almost all storage baskets are woven from rye straw and [are bound with] split stems of bramble or split branches of willow. Some baskets have handles and a lid. Farmers used them to store supplies and seeds; corn, peas, beans and other seeds. We also find oval sowing baskets in these museums. Some have a wooden handle, while others have a strap. When the farmer walked across the land, he carried the basket that had handles in his left arm, so that he could sow with his right hand. The basket with a strap was hung around the neck, so that the farmer could sow with both left and right hands.

The Open-Air Museum in Arnhem also has a bee house, in which various old beehives are exhibited. There is an old basket covered with cow dung from Huishorst (Gelderland), a cylindrical Boxmakorf, [named after its designer Teunis Boxma, (1863–1946)], from Assen, an angular Gravenhorster, or arch, from Arnhem (Fig 7), and a basket box from Ter Apel.

However, the ordinary beehive - which we have already seen from the Middle Ages onwards - is round in shape and tapered towards the top. Again, much variation is possible, depending on the traditions of the region or the taste of the weaver.

The material used for centuries by the beekeeper-weaver consists of straw or purple moor grass (*Molinia caerulea*) bound together by bramble stems or willow twigs.

Fig 7. A Boxmakorf (left) and a Gravenhorst Bow hive (right), Dutch Open-Air Museum, Arnhem.

In the seventeenth century, in an old bee book, *Den Naerstingen Byenhouder*, which is on measuring utility and profit, we can read the following:

"A Bee skep is a dwelling of the Bee, made here in Holland from rye straw, tied together with cleaved willow twigs, tapering from bottom to top, and somewhat flattened on the top. Rye straw is used because it is smoother and longer than wheat or oat straw, and because the bees have less to do to finish all the things inside before they begin their work [attaching comb] there. It is somewhat flattened at the top and tapering at the bottom, because it can be [inverted and] placed more easily with the open end upwards to allow inspection and management and they can fill the top of the hive sooner..."

1. OLD BASKETWORK

These reasons given for the traditional shape of the beehive are still valid today.

Lastly, we will look at some smaller household goods in the museums. These are baskets which were used in the kitchen or living room, baskets for storing eggs or peas or beans, potato peeling baskets, bread baskets and sewing or darning baskets. Until the beginning of the twentieth century there were many people living on small farms on sandy soil, where rye and *Molinia* grow. [In the UK, *Molinia* is usually found on wet, acid soil]

Some of them were also beekeepers with straw hives. To them we owe the old objects that are now in the museums or adorn our living rooms. Winter was the time for basket weaving, as the farmers were not able to work on the land at that time.

Moreover, in winter the corn was threshed, so it was not until then that rye straw was available. Bramble or willow, which were usually used for binding, were at the right stage of growth. [They were harvested as 6-foot lengths and then split lengthwise into 3 or 4]. The dried leaves of purple moor grass (*Molinia*) were also available in autumn for use as material for the coils.

Although we find many more baskets that were wicker (woven with willow twigs) appearing in print and paintings as well as in museums and collections, spiral basket work, [lipwork], is an ancient handicraft that dates back to the oldest times and which uses simple material found locally at hand and provided by nature.

BASKET WEAVING

2. MATERIALS AND TOOLS

In spiral braiding we need two types of material, on the one hand the <u>core</u> which is tied together in coils and on the other hand the <u>braiding,</u> [or stitching], band that binds the coils together. Rye straw or purple moor grass is usually used as the core and rattan or split blackberry stem is used as the braiding stitch.

Fig 8. On the left, purple moor grass (*Molinia caerulea*), and on the right, rye straw.

2.1 RYE STRAW

Of all types of straw, rye straw, the finer the better, is the most suitable material. Rye (see Fig 8 right) is usually grown on sandy soil, although not as much as in former times. Cereals are nowadays mowed by machine, threshed and pressed into bales. This happens, if the weather conditions have been favourable, in early August. Baled straw is less suitable, [unsuitable], for basket weaving. That is why the weaver should make arrangements with a farmer in good time to reserve some rye straw. Maybe it can be mown in the old-fashioned way with the scythe. Otherwise, it must be cut with a knife or shears. Rye straw is particularly suitable for our purpose because it is pliable and long. In addition, it has a golden - yellow sheen with many shades. Unfortunately, this shine disappears over a period of time. Other straw, such as that from wheat and oats, is stiffer and shorter and therefore less useful. When we want to store straw, it must first be dried. Damp straw will become mouldy and discolour. We also have to make sure that no rodents such as mice can get to it. We cannot immediately use the straw for making skeps or baskets. There is still some work to do in cleaning it up. We need to take off the ears and shake out any grass or weeds. The rye stems contain at least several nodes or growth plates, to which the long leaves are attached. These leaves are usually removed, although this is not absolutely necessary, as those leaves on the outside of the straw bundle can be removed when weaving. If they aren't removed, the skep looks rough and we wouldn't see the golden - yellow sheen so well. Of importance is that we cannot work with dry straw as it splits and breaks. It should be slightly damp so that it becomes smooth. Dry straw should be wetted, for example by putting it outside in damp weather, or sprinkling it with a flower spray an hour before. This applies to all types of straw. We can easily feel by hand whether the straw is flexible enough by checking that it no longer breaks.

2.2 PURPLE MOOR GRASS

Purple moor grass (Bunt grass or pipe straw) (*Molinia caerulea*) (Fig 8 left) is also called pipe straw because it was once used as a pipe cleaner. It grows [in The Netherlands] in large clumps on moist sandy and raised peat soils We find it in low heath fields and along sandy roadsides. Even though *Molinia* grows in the wild, we must still have permission from the landowner to cut it.

Sometimes it is cut to prevent the grassing of heather or the overgrowth of young tree plantings. In these cases, it may be easy to get permission. The best time to cut purple moor grass is the autumn or early winter. As an ordinary knife may quickly become dull when cutting, because of the [silica in the grass and soil], it may be better to use something with a saw blade. Furthermore, what we said about straw also applies here: purple moor grass needs to be dried before it can be stored. Unlike cereal straw the stems are not divided into nodes. The leaves are all together just above the stem base. This has the beneficial effect that all the leaves are loose if the purple moor grass is cut just above the ground. If we take a bundle at the top, beekeepers can simply shake out the leaves. Purple moor grass is a beautiful braiding material, although the colour is not as vibrant as straw. The thin, smooth stems are ideal for fine braiding work (see Fig 6). It is heavy, strong and durable, but also very stiff and often brittle. This means that a coil can easily break when making sharp bends, which mars the uniformly round shape of a basket. This can be prevented by moistening the bundle at a sharp bend or also squashing it by pressure. In general, it is not recommended that purple moor grass should be moistened before use, because it expands too much. Once the material has dried it would shrink and we would not be left with a sturdy basket. Because of its stiffness and difficulty of use, many weavers use a soaked bundle of hay or flax to make the round start of a basket.

2.3 RATTAN

Rattan comes from the climbing palm: rattan (*Calamus rotang*). Rattan grows in various tropical areas and can grow up to 150 m in length. The shiny outer skin of the rattan is our binding material, and it and the inner material are used to make rattan furniture. Rattan cane, (Fig 9 upper), is sold by weight in bunches of 500 to 1000 grams and can be of different widths. For our purpose, depending on the fine or coarser work we want to carry out, it should be from 3 to 5 mm wide. Sometimes this material is marketed as planed and edged. This means that the

Fig 9. Binding materials: above, split rattan cane : below, split bramble cane.

sharp edges and thickening of the knots have been removed. Sometimes it is available in craft materials stores, [or we may] have to rely on a beekeepers' suppliers.

When we have bought a bunch of rattan, we can start sorting by width and colour. In one and the same bunch of, for example, 5 mm, there are almost always wider and narrower pieces. And although the colour is natural, there are usually different shades in it. This sorting is not absolutely necessary. We can use the different colours and widths interchangeably. However, we must be wary of certain discolouration, because [this may indicate] poor quality. Where there are black spots, the rattan can easily break. Gray pieces may indicate previous mould and may now be so brittle that they break easily.

It is also recommended that rattan is soaked before use, for various reasons. Soaked rattan is less likely to cut the hand when it is being used. The main reason however is that when we have finished, a sturdy basket is created, because the rattan binding shrinks as it dries. Finally, the windings don't intersect when one touches the other while braiding. In short, rattan is a strong, durable and easy-to-use material. No wonder it has almost completely supplanted the so-laborious bramble and willow twigs of the past.

2.4 SPLIT BRAMBLE AND WILLOW TWIGS

Bramble, or blackberry, growing in the wild, can be found in almost all places in the Netherlands. It grows mainly on the edges of woods and along ditches and roadsides. There are many species and subspecies. Not every species is suitable for braiding. When we search for them in the autumn or early winter, we need to look for long, thickish stems about 2 m long without side branches. The stem should be green and slightly woody. Usually, the stems still bear half-withered leaves, but they are easy to remove. Depending on the thickness of the stem, 4-10 mm, after splitting we get narrow or wide braid bands (see (Fig 9 lower). In the past, a so-called 'puller' was used to cut blackberry stems, a stick with a sickle-shaped blade and a hook to pull the cut branch from the blackberry bush. Nowadays we use a thick cloth or gloves and pruning shears [or secateurs] because of the thorns.

After [cutting], various operations have to be carried out. In the past, the stalk was stripped of thorns by pulling it through a cow horn, in which a few holes had been drilled opposite each other. Now we can use a tube to scrape the thorns off the stem, the tube being slid around the stem and run up and down all sides when scraping (see Fig 10).

Fig 10. Cow horn and tube used for stripping bramble thorns.

After the stalk has been stripped of thorns, it has to be split into three and or four. This is not a task for a sharp knife. A traditional cleave, on the other hand, is much more convenient (see Fig 11). We can make a cleave ourselves from hardwood. When we are going to split, we first make three or four cuts in the underside of the bramble stem, into which the cleave fits. Holding the stem with one hand, we push the cleave through with the other. The stem splits quite easily and the parts diverge so far apart that there is enough room left for the hand pushing the cleave forward (see Fig 11). The split blackberry obtained at this point is not yet a braiding band. The next stage consists of scraping out

2. MATERIALS AND TOOLS

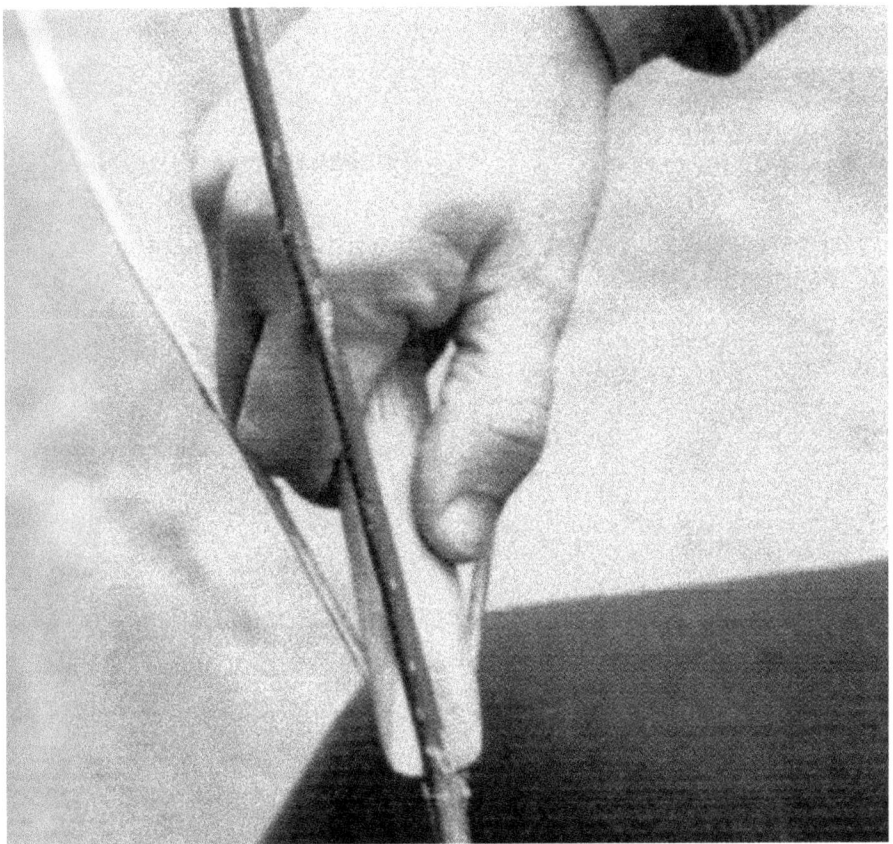

Fig 11. Using a cleave to split bramble.

the pith from the split stems with a knife. If some pith is left here and there, then the binding will slide a lot during braiding which makes it very difficult.

Finally, the split blackberry, especially when it's a bit woody, should be pulled around a round post to make it more flexible (see Fig 12). The cracking sound that we hear here means that the wood in the split blackberry is being twisted smoothly. It can be used straight away or can be hung to dry and then be stored (see Fig. 9). When we use it later, it must be soaked first.

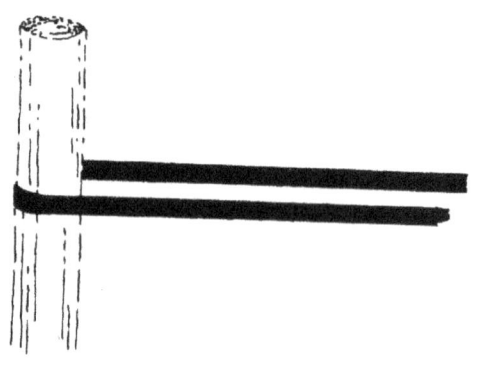

Fig 12. Pulling a bramble around a round post to make it more flexible.

It is thus clear that producing split bramble as a braiding material is very laborious. When braiding, it splits easily along its length, but otherwise it is strong. The lengths we get are usually shorter than rattan. That's why we need to turn in the end of an old one and the beginning of a new braid more often. This takes some time. For various reasons, we recommend that novice weavers use rattan. However, split bramble also has its advantages. If we don't mind spending time in our leisure activities, split bramble is a cheap material. But its main value is that its colour stands out more sharply against straw and purple moor grass than the lightly tinted rattan and this makes the braiding look so decorative. We can also use willow twigs for spiral braiding, as the 17th Century book *Den Naerstigen Byenhouder* tells us. The twigs are split in the same way as bramble. they must then be planed flat. and pulled smoothly on a round post. The same goes for split pine roots, which were formerly used for binding in Germany. Finally, we might mention rope, wire and coloured plastic ribbon, but these binding materials do not belong to the old and beautiful handicrafts. [Sisal string, a natural product, looks neat and works well when used as a binder for skep making]

2.5 TOOLS

We can be brief about tools, because they are simple. A needle or an awl is indispensable, (see Fig 13f). With this we make a hole

2. MATERIALS AND TOOLS

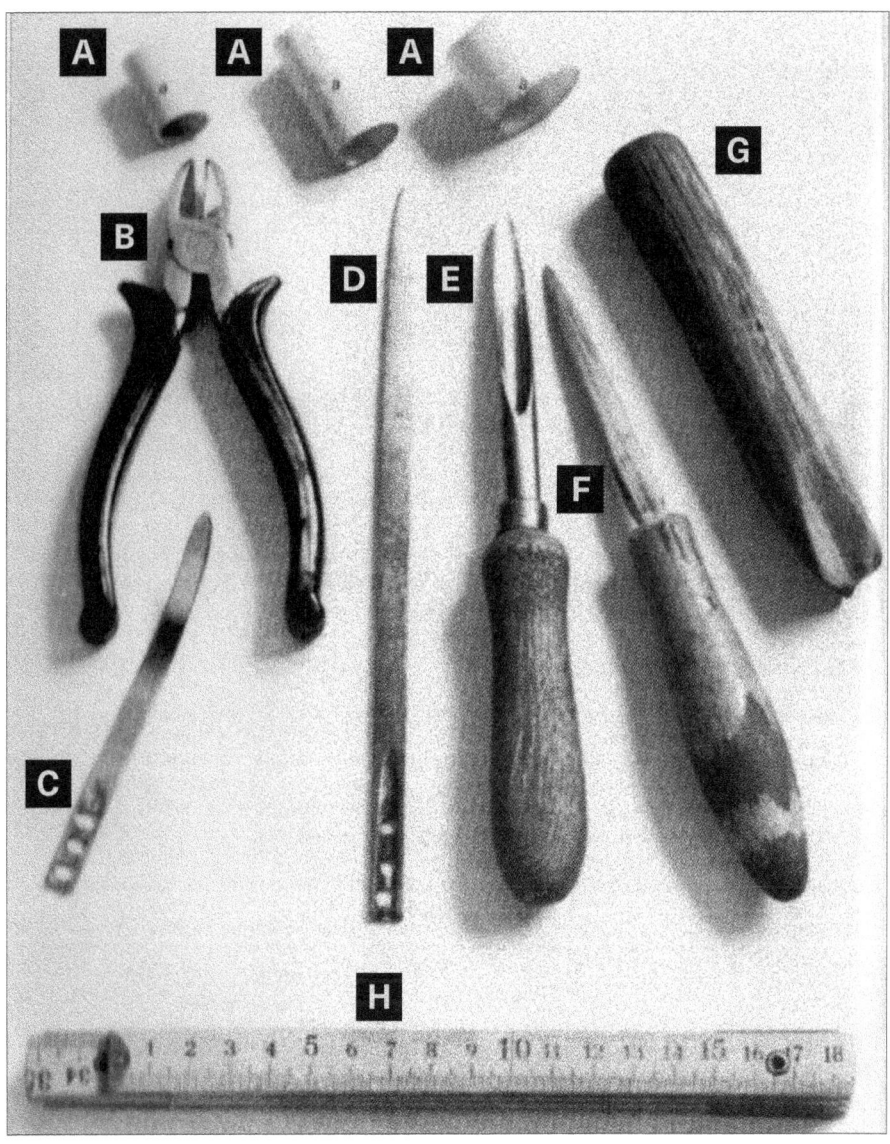

Fig 13. Tools:
A collars or braiding gauges
B wire cutters
C curved braiding needle
D straight braiding needle
E hollow awl
F wooden awl
G cleave for splitting bramble
H ruler

in the straw bundle, allowing us to pass the binding material as a stitch. In the past, an awl was sometimes made from a sheep or goat bone, because it was hard and smooth.

We can also make an awl ourselves from a piece of hardwood. When working with straw, a round awl (or a large nail) is used, but for purple moor grass, a flat awl is necessary. Some skep-makers work with a hollow awl [a fid] (see Fig 13e). This is a tapered tube that has been sawn along its length. The [fid] is put through the straw and the binding is placed in the hollow part and then goes through the space that the [fid] has created. Fig 13d is a flat braiding needle which can be bought from a bee-equipment suppliers. Flat needles can also serve as awls. At the base of the needle there are three eyes across its width. When we use rattan, we must first narrow the end of the cane and scrape it thin. That allows us to put this end through the first (top) eye of the needle then the third eye and finally back through the middle (see Fig 14). When the cane is attached like this, it is tightly fixed.

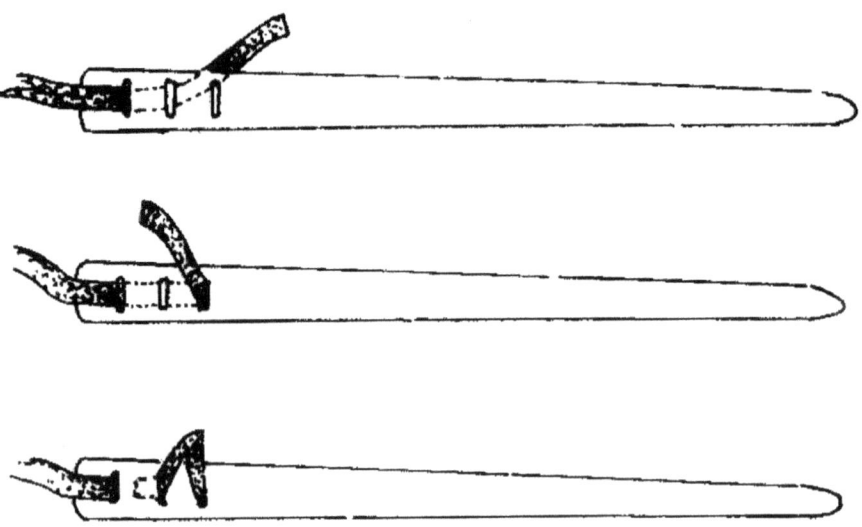

Fig 14. Needle and method of attaching the binding through the three eyes.

It makes a difference which end of the cane is fastened in the needle. The material of rattan is wire-like. We can feel or see, by the fibres of the underside and by the scales of the shiny top, how the thread runs. We have to fasten the cane so that it moves with the thread when braiding, otherwise it will fray. When using split bramble, we have to make two punctures, once with the awl and then with the bramble length through the hole made by the awl.

In addition to these tools, we also use a so-called collar [or gauge]. This is just a tube that is slid around the bundle of straw to keep it at the same thickness. A piece of cow horn or bone used to be used for this, because the slight taper made it easier to insert straw. A gauge can be easily made of leather, wood or metal. A piece of aluminium or plastic will do just fine. [Plastic bottle tops make easily available and effective gauges]. So, this does not have to be bought specially. How wide a collar should be used depends on the thickness of the bundles of straw with which we want to make our skep or basket.

BASKET WEAVING

3. THE TECHNIQUE OF BASKET WEAVING

With the help of working drawings we will try to clarify the technique of spiral braiding or stitching. The instructions, [which are for right handers, left handers should reverse the direction], must be followed carefully, otherwise the weaver will face unpleasant surprises. For the sake of convenience, we will speak of straw, though it might be purple moor grass or some other material which might require special handling.

3.1 START OF DISC AND BOWL

We take a bundle of straw [about 10 or 12 straws]. If the straw is not flexible, it should be moistened. Purple moor grass can be crushed or bruised to make the first sharp curves more easily. Then we take a piece of stitching material of about two metres. We hold the beginning of the stitching material and the bundle with thumb and forefinger at the same time and wrap the stitch around it several times (see Fig 15/16 a and b). The stitching material should be pulled tight. In this way the stitch fixes its own beginning and at the same time the straw is tied together into a sturdy bundle. The first round is an eye (see Fig 15/16 c), [with only a small gap in the middle]. The flexible bundle is still thin, but it will soon have to become thicker [between Fig 15/16 c and Fig 15/16 d]. Then we can use the collar or gauge [to set a fixed width]. The second round is tied to the first by stitching through the eye several times (Fig 15/16 d)

BASKET WEAVING

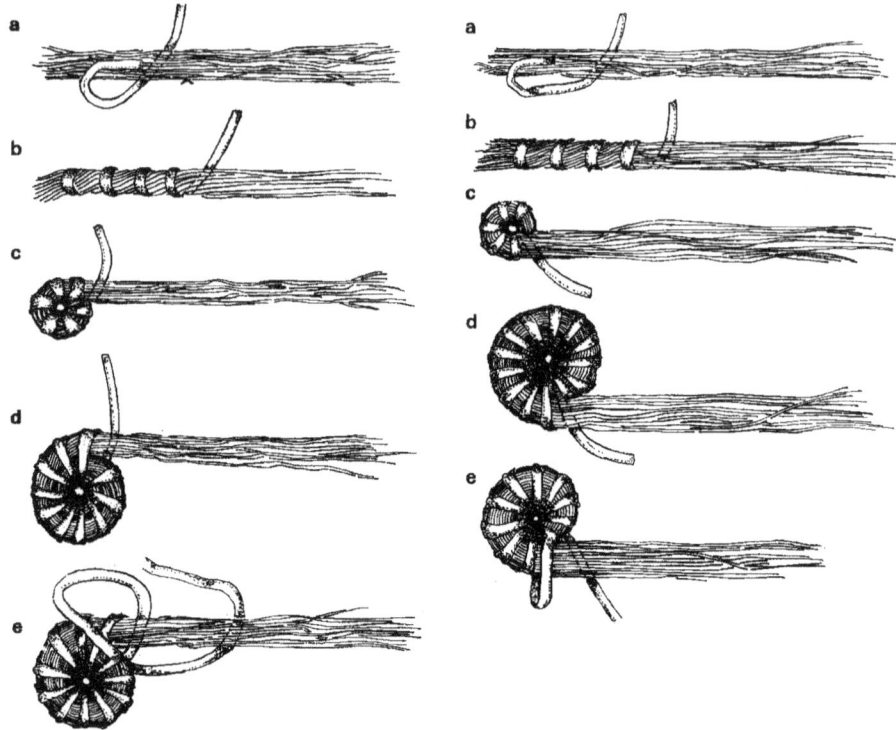

Fig 15. Five steps to get started.   **Fig 16.** Five steps to get started (variation on Fig 15).

3. THE TECHNIQUE OF BASKET WEAVING

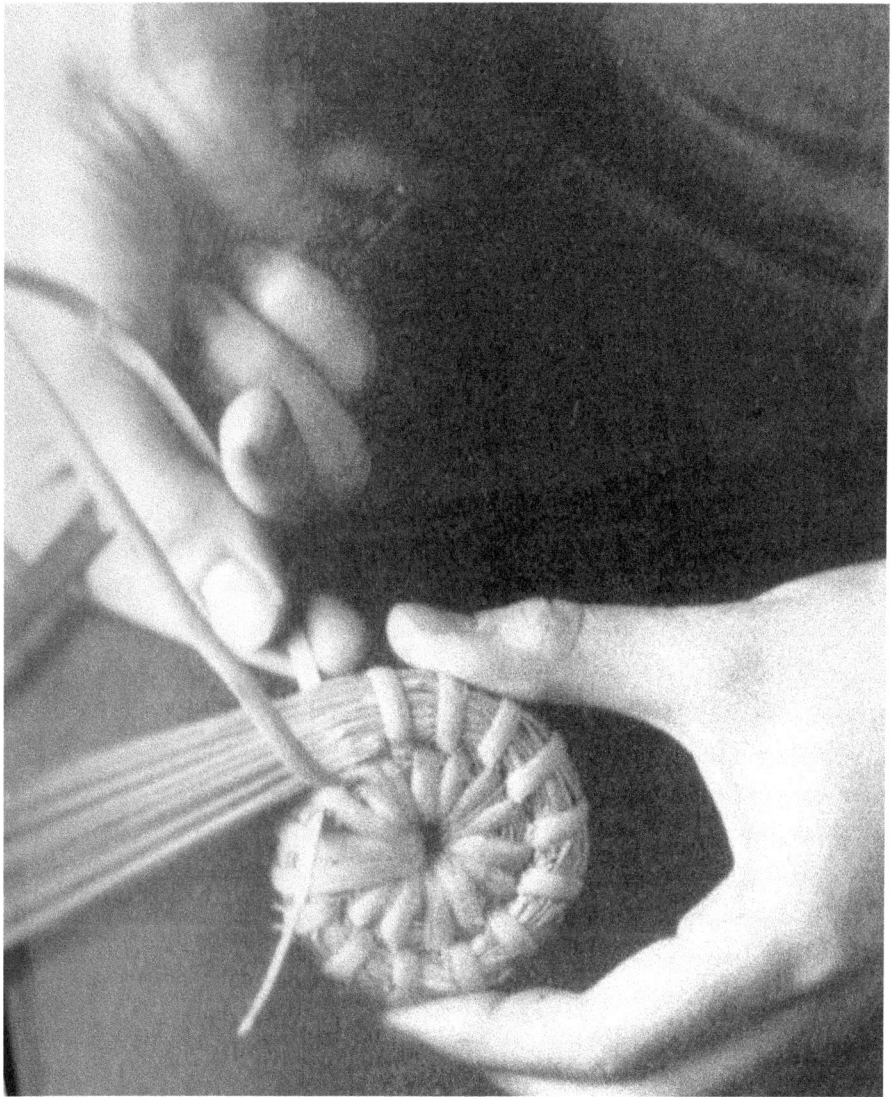

Fig 17. The third round.

The smaller the eye is the better, otherwise a hole will be left in the centre of the dish or basket. After the first two rounds are tied together, we need to use the needle or awl (see Fig 17). At the start of the third round, we no longer put the stitch through the eye, but through the underlying [second] coil. We do not place it between the coils but slightly below this, through the lower coil

BASKET WEAVING

so that part of this coil is incorporated in the stitch (see Fig 15e). [The stitch goes diagonally across the previous stitch] We do the same with every stitch that follows.

There is another way to start that is often used when weaving skeps. We take a fairly thick bundle of straw and tie the stitching material around it with a clove hitch. This knot is placed at a distance of approximately 8 cm from the beginning of the bundle (see Fig 18a).

Fig. 18. Another way to start.

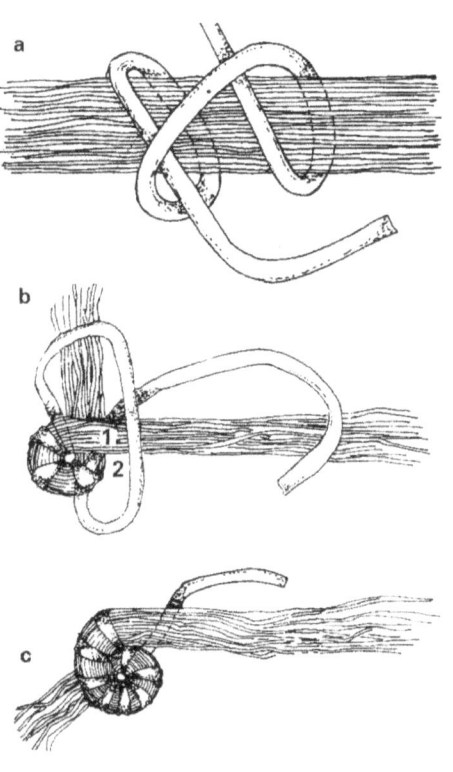

We pull the clove hitch tight and then the bundle is twisted spirally around the starting piece at the place where the knot is located. This twist is the first round. To tie it up, we wrap the stitching material around the twist. Then we pass the stitch under the winding of the knot and repeat this until the first round is tied (see Fig 18b). When the first round has been plaited like this, the short end of the bundle sticks out (see Fig 18c). This end must be cut away afterwards.

Furthermore, we place each new stitch under the already applied stitches from the first round. This way, our stitching will not leave a hole in the centre (see Fig 19). If this first round is a bit lumpy, we can knock it flat with a hammer.

3. THE TECHNIQUE OF BASKET WEAVING

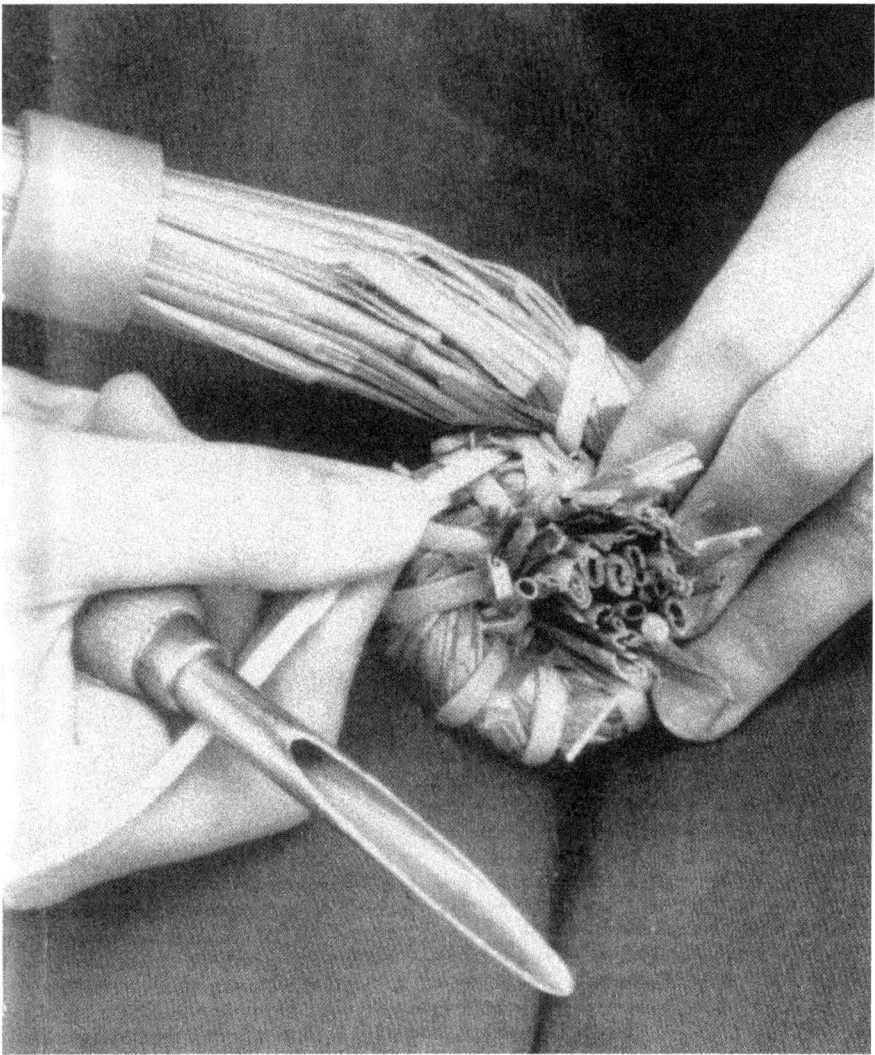

Fig 19. The beginning using the twist.

3.2 SUBSEQUENT ROUNDS

Before considering the making of subsequent rounds, we need to address three new issues. These are a) the need to add in extra stitches, b) the need to add straw and c) we must learn how to secure the end of one length of braiding material and the beginning of a new one.

Adding extra stitches

This must be done soon, as the coils quickly increase in size, especially early on. If the stitches are too far apart, the skep will not be firm. Purple moor grass doesn't coil neatly unless there are sufficient stitches and this can detract from the evenly round shape. To add an extra stitch simply put it through the straw between the two underlying stitches (see Fig 20, where there is an extra stitch between a) and b). It is also possible to make an extra stitch without passing the stitch through the straw stitches, but just around the bundle of straw. Whenever the underlying stitches are too far apart, we have to add extra stitches. In fine work, stitches should be proportionately closer together than in a large, coarse basket. A common measure of the distance between the stitches is the thickness of the straw bundle or slightly less.

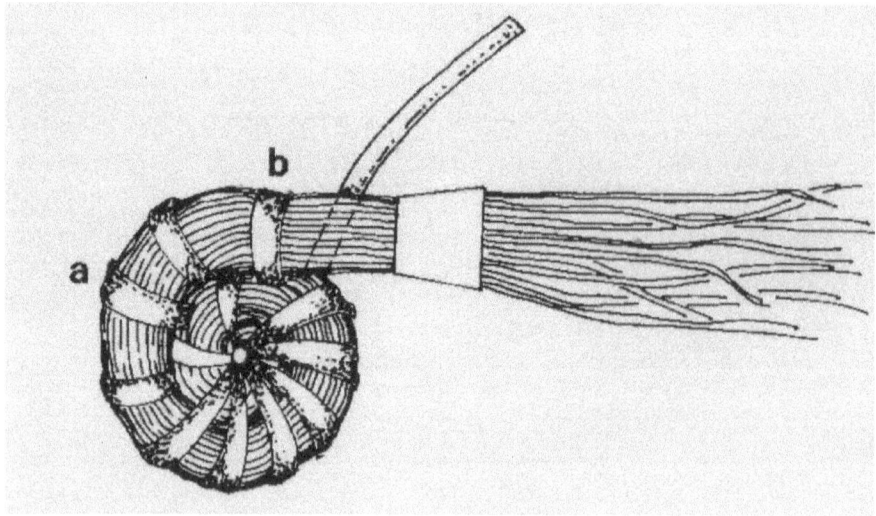

Fig 20. Adding stitches.

Adding more straw

The easiest way to start is to use a thin bundle of straw. After the second round we have to add straw to get the bundle to the right thickness. Before we do this, we slide the collar or gauge around

3. THE TECHNIQUE OF BASKET WEAVING

and down the straw until it is a few cm away from the last stitch. Now that we're going to give directions for adding straw, we need to distinguish between straw and purple moor grass as there are differences. With straw, we add it to the inside of the bundle until the collar or gauge feels full. This must, of course, be repeated as you continue. If the collar becomes loose then that's the signal to add more straw.

As the plaiting with straw continues, a special point must also be mentioned here. Before tightening a stitch or several stitches, we can twist the straw through half a turn.

In this way the hollow straws are squeezed shut and a sturdier straw bundle is created.

[However, if the straw is left untwisted, the air in the hollow straws enhances insulation].

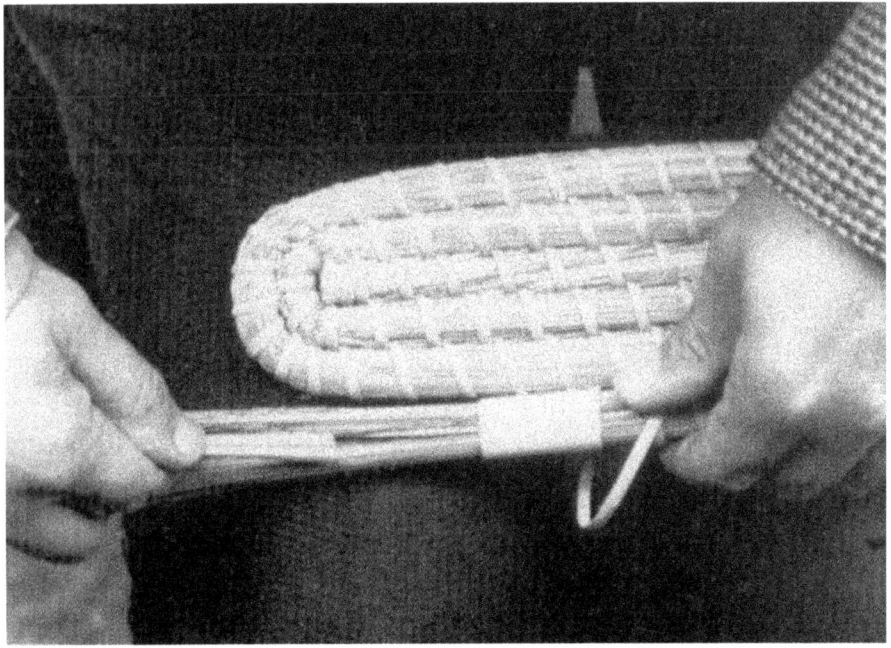

Fig 21. Adding purple moor grass.

Purple moor grass can be inserted further into the collar (see Fig 21). Since it is stiffer than straw, we can insert it through the last tightened stitch. Care must be taken that no loose grass or straw comes out of the bundle anywhere, as this disfigures the finished appearance.

Connecting two lengths of stitching material

A manageable length of stitching material is about two metres. When little of this length remains, a special skill is required. We leave the end of the old stitching material, with which we made the last stitch, at the back of the bundle of straw (see Fig 22 a). We insert the new length of stitching material through the hole of the last stitch and pull it all the way through, except for a little bit.

Fig 22. End of the old stitch (a) and beginning of the new one (b).

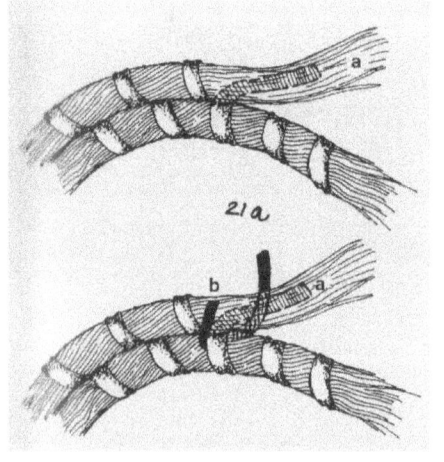

We bend the short end to the right when we have the disc in front of us (see Fig 22b). Then the new stitch is put under the next coil, as we have done before. This stitch simultaneously covers the short end of the old stitch at the back, the bundle of straw and the short end of the beginning of the new stitch. We can put these ends on the outside, but it's much neater to push them both in the middle under the straw bundle. With the first new stitch, its short end is immediately secured, because there is a sharp fold in it. The end of the old stitch, on the other hand, has to be tightened again when making the next stitch. If

3. THE TECHNIQUE OF BASKET WEAVING

the end is too long, we can cut it off. There are several different ways to tackle the beginning of a new stitch and the end of the old stitch.

A variation on the previous method is where we don't run the new length of stitching material through the hole under the last stitch of the old length.

Instead, we place it in the middle of the lower coil, pass the straw bundle under the last stitch of the old length in such a way that it comes out under the end of that stitch (see Fig 23). As a result, the start of the new stitch is invisible and the end of the old stitch can easily be pulled in under the straw bundle.

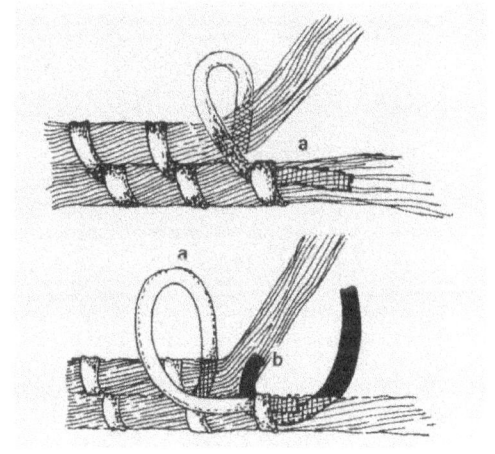

Fig 23. The start of the new stitch: a is the end of the old one, b is the beginning of the new stitch.

Another easy and frequently used technique consists in connecting the end of the old length of stitching material with the beginning of the new one. When we see that we're nearly at the end of the old length we place the start of the new length, inner side up, under the last three stitches of the round we are braiding (see Fig 24). Then we put the end of the old stitch under the new one with both approximately in the middle of the straw bundle (new on top of the old). At the point where the old and new lengths meet, we take them between thumb and forefinger and give them half a turn towards us. Furthermore, we have to make sure that the end of the old band is on the top right of the straw bundle, we wrap it when we continue with the new stitch (see Fig 24 and 25). The new stitch to be inserted is laid with the inner (rough) side up for a good reason: after a half turn it is back into the right way up. This stitch should not be used on the last round of the finished product, because it doesn't look neat. In these circumstances we should put the start of the new stitch into the underside of the bundle and tie it in there, or we can use one of the techniques mentioned above.

Fig 24. Attachment and securing of the old and new stitches.

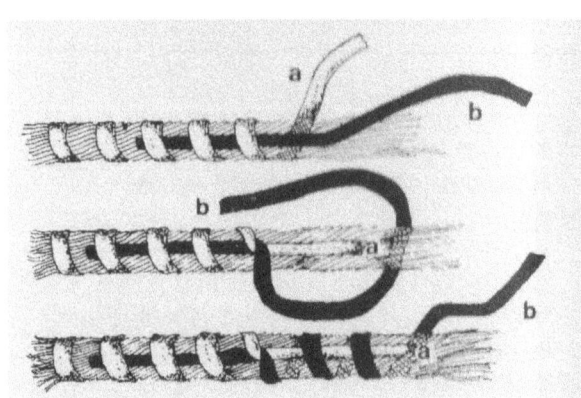

Fig 25. Joining: (a) is the short end of the old stitch, (b) is the beginning of the new stitch.

3. THE TECHNIQUE OF BASKET WEAVING

Fig 26. Disc and bowls.

Fig 27. Inserting the stitching needle.

[A stitch can be inserted and then tightened, or several stitches can be inserted, without initially tightening each one. (See Fig 27 above). Once several stitches have been inserted like this, they can then be tightened one by one, before several more stitches are inserted]

3. THE TECHNIQUE OF BASKET WEAVING

Fig 28. Pulling the stitch through.

3.3 DISCS AND BOWLS

As we work away, a disc forms. From the beginning we have assumed in the directions and drawings that we are right-handed. Left-handers can easily perform the necessary actions in reverse. If we hold the disc in front of us, we have to add straw from the right. Each new stitch is inserted in the side facing us (see Fig 27). We also pull the stitches tight at the back with the right hand (see Fig 28). When tightening, it's a good idea to use the thumb and forefinger of the left hand to hold the previous stitches or to press against the sides of the newly laid stitches that we want to tighten a little more.

It is also possible to insert the stitches at the back of the disc. We should imagine this as follows. We again hold the disc in front of us, while the bundle of straw is below. Then we insert the fid or needle at the back and pull the stitch towards us. When we work like this, we have to be aware of this from the beginning (see Fig 16). Once the disc is big enough for our purpose, it needs to be brought to a finish. The straw bundle should never be cut off.

If we stop adding straw, the size of the bundle decreases. We may even occasionally pull straw out from the bundle. The collar or gauge is no longer needed. The final coil becomes thinner and thinner and eventually runs out completely. Reducing from the full coil width to the end should occur in about a quarter of the circumference of the disc. We can safely cut off the last 3 straws. At this point we put several stitches tightly against each other a few times (see Fig 29). Finally, we bury the remaining cane at the back and cut it off.

Fig 29. Finishing off.

For our use, a disc can serve as a coaster or as a decorative container for a dried bouquet (see Fig 26). How do we make a disc into a bowl? With the disc we already have a bottom, it is the raised edge that is missing. To raise the edge, we hold the disc in front of us and the bundle of straw above is placed diagonally downwards (see Fig 30). Because of these stitches, the bundle ends up half on top of the outer edge of the disc.

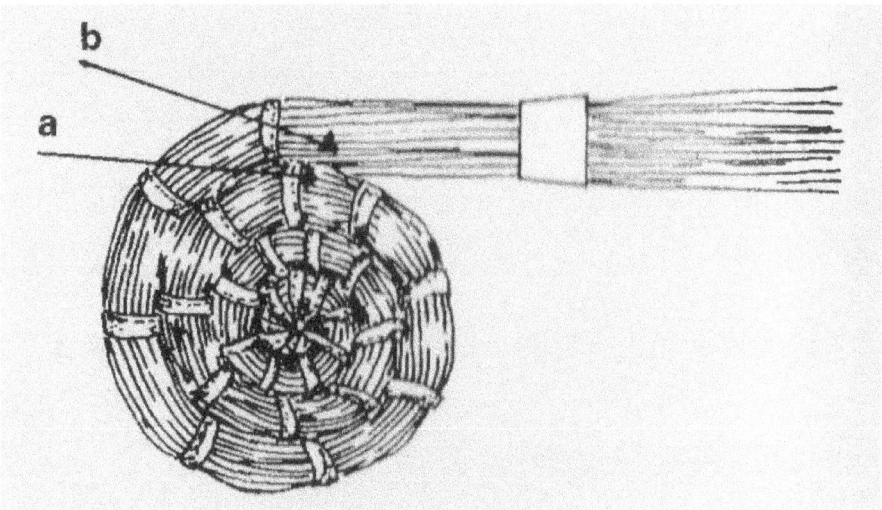

Fig 30. a) is the direction of the bundle before raising the edge b) is the new direction of the bundle.

When we stitch in this way three or four times, a low shell is formed with an outwardly extending rim (see Fig 26). When we do this, straw behaves differently from purple moor grass. As we tighten the stitch on the inside of the shell, straw is more flexible than the purple moor grass. In order to get the straw bundle in the right place, we must insert the stitch less obliquely.

3.4 DESIGN PRINCIPLES

We have already discussed an important design principle with the outwardly extending ridge. If we make the bowl higher, a basket is created.

We can also weave a bowl with a straight upturned edge, ie one that is cylindrical. Thus, starting from the disc that we hold flat in front of us, we must insert the stitch horizontally. We cannot simply switch from vertical to horizontal. The sudden transition is not attractive. We need to make some transition stitches. As soon as we gradually insert the binding material horizontally, the bundle of straw forms the first round of the straight raised edge (see Fig 31).

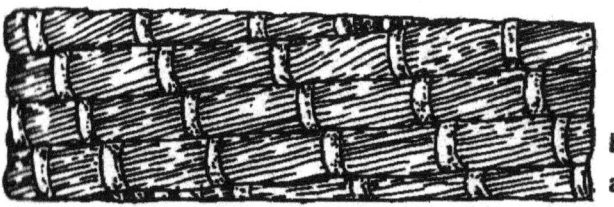

Fig 31. Round b is placed on round a. The first stitches of round b are transitional.

A third possibility is that we create bowls with an inward edge, a strange model. To do this, we first have to make a round with a straight upstanding edge on the disc and put it diagonally inwards in the succeeding rounds. We mention this possibility because it is also an important principle for the design of baskets. With these three types of stitches we can determine the shape of what we are making. Using them in combination, depending on our imagination, beautiful shapes can be created. Fig 32 shows a basket in which the three possibilities are combined. The lower part of the basket extends outwards, the middle part is straight up and the upper part slopes inwards; all this in a smooth line resulting from the three ways or directions of insertion.

3. THE TECHNIQUE OF BASKET WEAVING

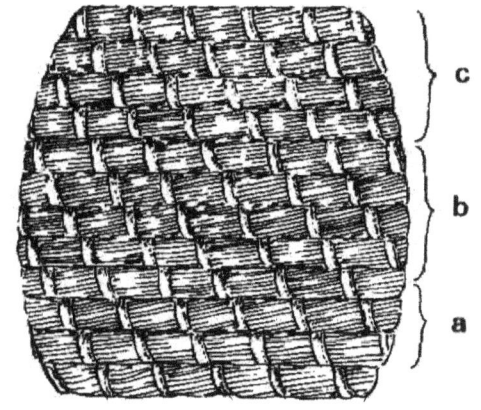

Fig 32. A curved shape, created by three different insertion methods a) the lower going outwards, b) the middle going straight upwards, c) the upper coming inwards.

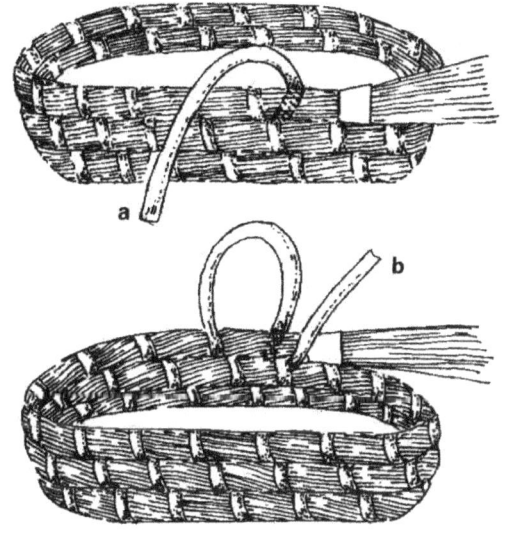

Fig 33. Two possibilities, for which the same rules apply: In both, straw is added from the right a) the cane is inserted from the outside, b) the stitches are tightened on the inside.

It may be best to repeat an important point. When weaving dishes and baskets, round or oval, it is vital from the start to insert the weaving stitch from the outside (see Fig 33). Our hand does not have enough room to manoeuvre the needle or awl in the inside of the basket. If we make a small basket, we can only use the awl from the outside, because there is not enough room on the inside to put the needle or awl through the straw bundle.

3.5 CORRECTING ERRORS

When the stitching material becomes twisted, we need to take out the stitch and reinsert it. If we stitch with a needle, this also means that we have to remove the material from the eye of the needle and reattach it. Sometimes we find that we have skipped an underlying stitch. If we discover this very late, we probably won't be able to muster the courage to re-do much of our weaving. Pulling out straw isn't very helpful. So, just ignore the mistake.

We often don't notice until quite late that our skep isn't symmetrical. If this is only to a small extent it makes no practical difference. The reason is that we have placed part of a round too far out, and every successive round will have to follow its predecessor.

When we see this asymmetry at a late stage there is nothing that can be done. If we discover the mistake on the round just completed, then we can take it out or adjust on the next round. In this round we have to put the straw bundle, depending on the mistake, slightly more outwards or inwards. This is how the fault may be made less conspicuous. In a bundle of purple moor grass a bend can occur, with straw it is less likely. When purple moor grass is not made sufficiently flexible, this bend easily occurs when the rounds being made are small. It is then advisable to take out a stitch and to make another stitch to cover the bend. Thus, the fault is somewhat restored.

3.6 MAKING AN OVAL SHAPE

When making an oval shape we need a different start from that used for a disc or bowl. We don't need all the force with thumb and index finger involved in making the first small round. With the oval shape, we can immediately start with a thick bundle of straw. The thickness should remain the same throughout.

3. THE TECHNIQUE OF BASKET WEAVING

We wrap a bundle of straw over a certain length. This length depends on the size of the basket we want to make. The beginning of the stitching, which we have to put in the bundle, sometimes does not stay in place. To prevent this we can tie a knot in it. The beginning of the bundle should be nice and smooth or we should cut it straight before the second round starts. Then we make such a sharp bend, just past the wrapped part, that the bundle is doubled.

Also, in the bend we have to put in some stitches. Finally, we have to pass the new stitches under the previous stitches, [see Fig 34] to allow us to lock in the next line of stitching. Straw manages these tight bends without much problem. Purple moor grass, on the other hand, needs to be crushed and moistened at the bend. This is how we deal with sharp bends over and over again.

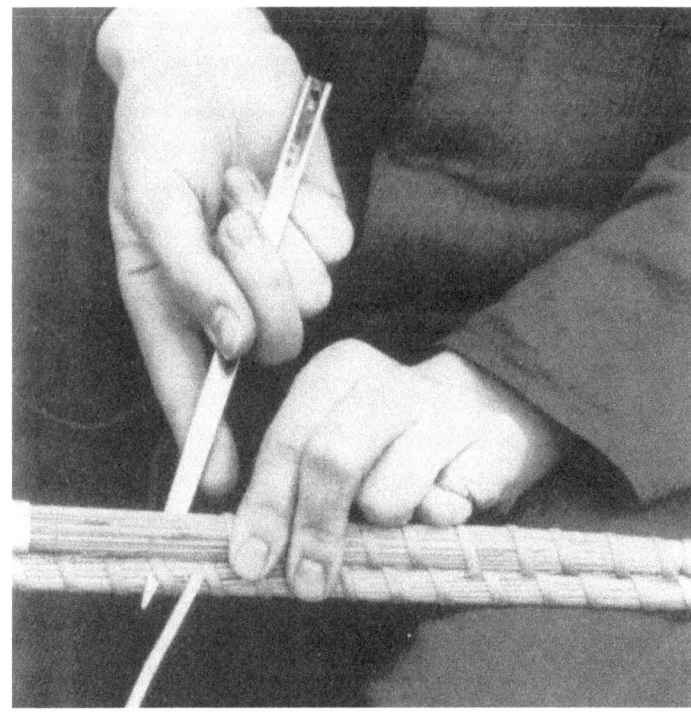

Fig 34. Beginning of the oval base.

We also have to remember that crushed purple moor grass becomes thinner when it is curled in a coil. This happens because, by crushing, the central air space disappears from the straws. To keep the bundle of the same thickness in the bends, we can put a piece of [willow or hazel] in it.

A further problem that can arise with the oval base of purple moor grass or straw is that the bottom can become skewed. In this case, before starting the first round of the edging, we can nail the base between two pieces of wood (see Fig 35). [One piece of wood covers the inside of the base and is then nailed to a larger piece placed below]

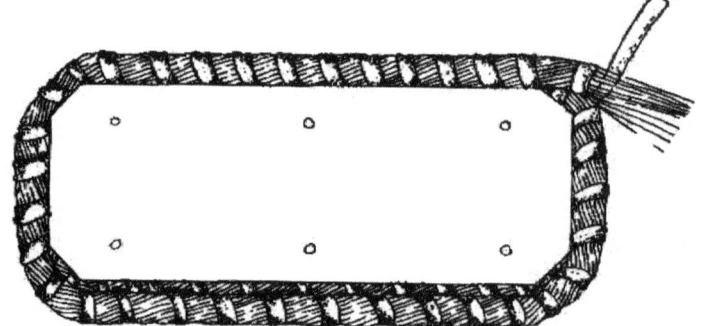

Fig 35. Planks of wood used to prevent skewing of an oval base.

The outer round of the base must remain free, because on this we have to stitch the raised edge. After stitching a few solid rounds, we can loosen the wood and the bottom remains flat. We can of course make a low or high shape with spiral coiling (see Fig 36).

3. THE TECHNIQUE OF BASKET WEAVING

Fig 36. Oval baskets.

In addition to the oval shape, we would like to say something about the angular shape. With straw we are able to create 'round corners', so that an angular basket is created (see Fig 7). The start is the same as with the oval bottom. In the sharp bends we then have to twist the straw in such a way that the bundle makes an angle. In fact, the material only just allows this manipulation. It is even more difficult with purple moor grass, which is much stiffer, but it is still possible (see Fig 37). Such a basket with crushed or bent corners will wear out more quickly.

Fig 37. Angular basket with a flat bottom.

3.7 CURVED SHAPE

For lack of a clear name, we have called this the curved shape (see Fig 57). It is the most common form of the *lectuurmand*, [a basket for holding reading material, newspapers, magazines, etc]., but more-or-less curved shapes can also be woven for other purposes. This form is created by playing with and applying the principles of design. We start the shape with a large disc. Then we have to weave so that a raised edge forms on two sides, while the other two remain more or less flat. When we imagine the disc lying flat in front of us, the following applies: where we want to make the raised edges, we put the bundle of straw *on top of* the last round; on the other two, we place the bundle *alongside* the last round. We have said before that the transition from side to side should not be made suddenly. In order to get a curved shape, at least a quarter must consist of transitional stitches. So, in each round we have to put stitches into the straw bundle from different directions. In order to make this easier to understand we can divide the disc into quarters and mark this with nails (see Fig 38). After one or two rounds we need to move the nails further to the side. In this way we continue until the curve has reached its planned position.

3. THE TECHNIQUE OF BASKET WEAVING

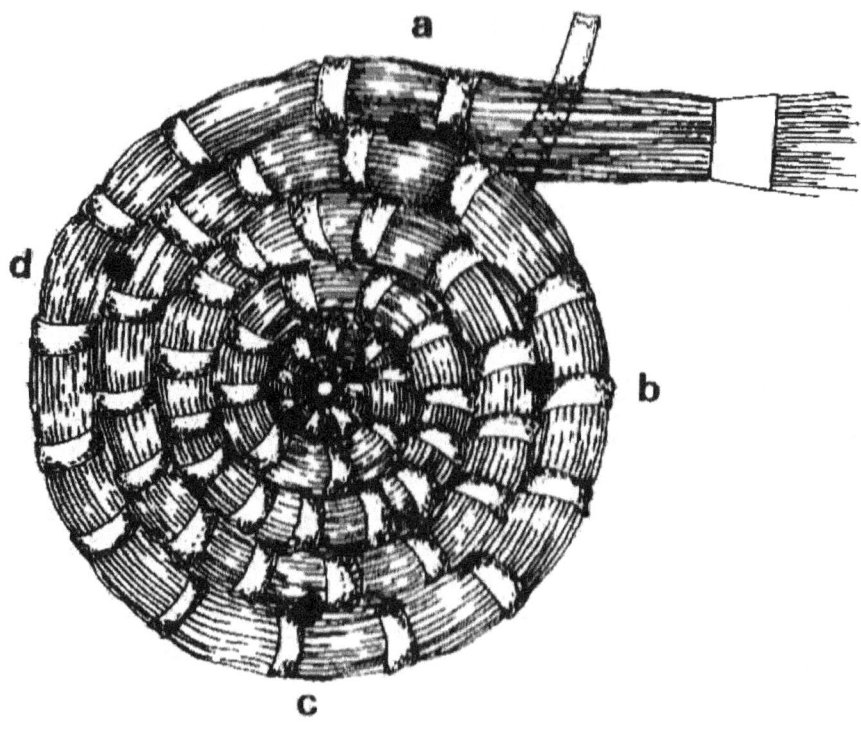

Fig 38. Beginning of the curved shape. The raised edges form around b and d, while a and c enlarge the flat bottom. From a to b the bundle of straw gradually rises and from b to c it lies gradually closer to the previous round. The same goes for the other half round from c to d and from d to a. Accordingly, insertion should gradually transition from vertical to horizontal every fourth part of the round.

Fig 39. The Apiary at the Ambrosiushoeve at Hilvarenbeek.

4. BEE-SKEPS

The bee-skep represents an ancient form of spiral weaving. It appears repeatedly in medieval art (see Fig 3). Although the shape has changed little since those times, it was much taller then. Nowadays, the moveable frame hive has almost completely replaced the skep. Because the bees attach their combs to the inside wall of the skep the beekeeper had to kill the bees to break out the combs in order to harvest the honey. [An alternative, was to "drive" the bees upwards into an empty skep, separating the colony from the cleared combs below] In a modern beehive the combs are drawn on moveable frames, which the beekeeper can take out whenever he wants, extract the honey and then replace the frame. For this and other practical reasons, modern hive beekeeping has replaced skep beekeeping. Yet there are reasons why every beekeeper should still have a skep available. Whatever method of swarm prevention is used, a swarm will occasionally occur. Nothing is more convenient than a skep, into which the swarm can be collected. The skep can be a normal tall one or, even better, a shorter, wider one [to make collecting a swarm easier.]

Furthermore, the skep is more practical than a modern hive, as it allows interested visitors to admire the bees making comb, The beekeeper only has to pick up the skep and hold it an angle and the bees are hardly disturbed. [This is best done with the skep held so that the combs are vertical, for stability]

Another comment, on the sustainability of skeps. The material is very perishable, especially when it is populated by bees and exposed to wind and weather. The beautiful appearance fades after a few years. In the past and still sometimes today, to prevent wear and tear, the skep was treated with a mixture of loam or clay and cow dung to increase its durability. [A process known as clooming].

When we come to discuss making skeps, we are not concerned with unusual models such as the box hive (see Fig 7), or others illustrated in Fig 39. We will limit ourselves to a normal skep and give instructions to make it ready for operation. For beekeeping with skeps, you should consult the excellent manual by JJ Speelziek. *Korfimkerij (Skep Beekeeping)*

4.1 DIMENSIONS AND ENTRANCE

Skeps can be with flat or with pointed tops or they may be intermediate between these forms (see Fig 40). A slightly flat top is preferred so that the skep can be placed upside down when the bees are being put into the skep. As far as the material is concerned, rye straw is most suitable for retaining heat in the skep, but purple moor grass or other straw can also be used.

The weaving technique for making a skep poses few problems if we have learnt how to make a disc and a bowl. As mentioned previously, a common way of starting is to use the alternative method described earlier (see Fig 18). We need to make sure that the remaining length of the straw bundle is on the inside of the basket. Then, after the first few rounds, we can cut it off. Once we've completed the beginning, we can close any hole in the top of the skep, [if one was left at the start].

The skep has a height of 40-50 cm and is about 33 cm in diameter. The first part of the skep, the top section, is nothing more than a shallow basket that has not yet been completed. To continue, if the dish has a diameter of 25 cm, for example, then we know that the following rounds of the basket must always be positioned slightly outwards, until the diameter reaches 30 cm. Reaching the required height is, of course, dependent on the thickness of the

4. BEE-SKEPS

Fig 40. Various skeps. At centre left is a swarm-catching skep.

bundle of straw. At the top of the skep, the bundle is still rather thin. Once the top has been completed it will take about 15 rounds of 3 cm thickness to complete the finished skep. When we work, not with purple moor grass, but with straw, the diameter of the collar does not indicate the thickness of the straw round. After all, straw is twisted or pressed together between the collar and the stitches.

By the time the ninth round is reached, (about three fifths of the skep), we have to make the entrance. A higher placed entrance is not recommended. In winter, the bees do not like to be positioned in front of an open door. They often seal it themselves. The entrance can be made in the following way.

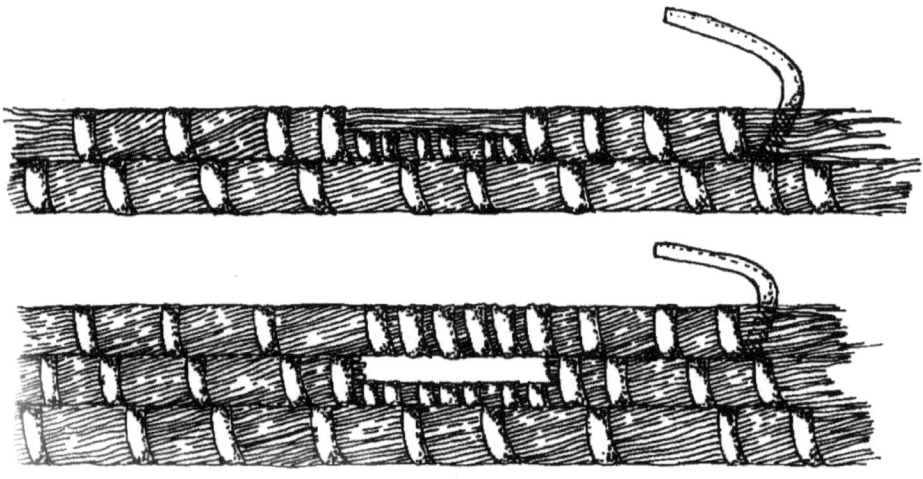

Fig 41. Making an entrance.

In the place where it is to be, we first put in a double stitch. Then, we don't put stitches *around* the coil being added, but instead place them from the inside of the skep *through the centre* of the coil. Then we make several stiches like this side-by side going through the underlying coil and half way through the overlying coil. As soon as the entrance is about 8 cm long, we make another full wrap around the whole coil and continue as before. (See Fig 41). These close-lying stitches serve to reinforce the entrance. When we have done a few more stitches, we need to cut away the unstitched straw. This gives an entrance about 1.5 cm high. After completing a round, we come again to the entrance. At this point, we need to insert several closely placed stitches in the coil. This forms the base of the entrance (see Fig 41). [An alternative to the entrance described above, is to place the skep on a wooden or stone base which has a built-in entrance.]

Once the skep is tall enough, we let the straw bundle get less until it runs out. For this, the bundle must be placed flat on the previous round, possibly by inserting it lower. For reasons of appearance, we prefer to complete the last round opposite the entrance by means of a few extra stitches if necessary.

Sometimes the beekeeper uses the stitches to make a kind of decorative edge on the last round of the skep. This decorative edge is actually intended to reinforce the bottom edge, on which the skep rests and where it wears out most easily, (see Chapter 6 on Embellishments).

We can be brief about the shallower swarm catching skep that we mentioned. It is about 30 cm high and 40 cm wide (see Fig 40). Its width makes it easier to catch a broad hanging swarm. This skep usually has no entrance, but it may have a handle on its fairly-flat head (see Chapter 5 on Handles). The flatter top makes it easier to identify the queen when we have a swarm, as the bees are spread over a larger area.

4.2 AN EXTRA BASE FOR THE MAIN SKEP

If, in the opinion of the beekeeper, the bees do not have enough space in the skep, he can make an extra base [a nadir, which goes below the skep]. This structure, made up of perhaps four coils, is difficult to weave freehand. After the beekeeper has measured the size of the skep, he can draw a corresponding circle on a board of wood. Around the circle he hammers a number of nails that should protrude through to the other side of the board. He pins the first round of the base's edge onto this while stitching (see Fig 42). When the extra base is ready it must be attached to the skep with hooks.

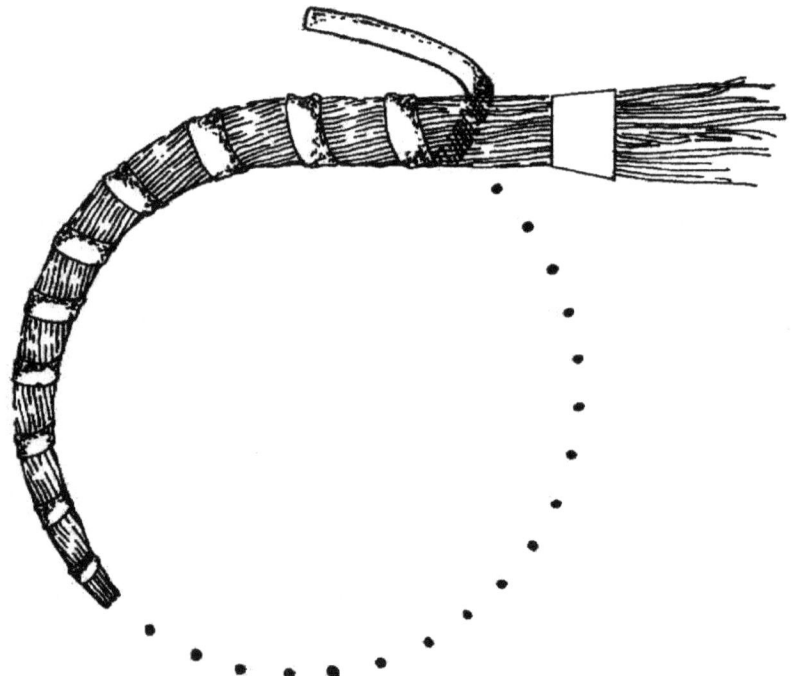

Fig 42. Starting the extra base on a plank using nails [or golf tees placed into drilled holes].

4.3 BARS FOR THE SKEP

The skep is not ready for use until it has been fitted with bars. [Also known as spleets] These [are slats or supports for the combs which are] inserted into the skep., The bars, about 1 cm thick, can be round or triangular. [One end is sharpened to a point and the other end is blunt]. The question is how to insert them. The beekeeper generally finds that the bees produce their combs in what is called the "cold way". With this construction the combs are attached to the front and back of the skep. The beekeeper can more-or-less force the bees into producing their combs the cold way by "arrowing". We take the entrance as a starting point for the "arrows". We push [the sharp end of] a bar through the back of the skep, so that the point sits just below the entrance. The bar is placed horizontally and is inserted through the body of the straw round, not between two straw rounds.

The blunt end must not be fully inserted into the straw round, [because] when we wish to remove combs later on, we need to be able to grasp the blunt end to pull the bar out with pliers. Next, two bars, horizontally and parallel with each other, must be placed crosswise above and below the bar which is just below the entrance.

Each pair of bars should divide the horizontal section of the skep into approximately three equal parts (see Fig 43). The so-called header bars deserve separate consideration. These are three rather short bars which are inserted near the flat top of the skep. They are inserted in the same direction as the bar which has been put in near the entrance.

From these header bars the bees begin to draw combs in the cold way. If the header bars are 10 mm wide, they should be 27 mm apart, because the bees need 37 mm for the width of the comb and the corresponding bee space. To make it more likely that the

BASKET WEAVING

Fig 43. A Skep with bars inserted and a decorative edge.

4. BEE-SKEPS

Fig 44. Skep with bees (Ambrosiushoeve in Hilvarenbeek).

bees will build comb in the cold way, it is a good idea to attach pieces of comb to the end of the bars as the bees like to use these as starting points. This is the tried and tested way (see Fig 45), but other ways, may, depending on local conditions, be just as good (see Speelzeik, *Korfimkerij* pp 61-62).

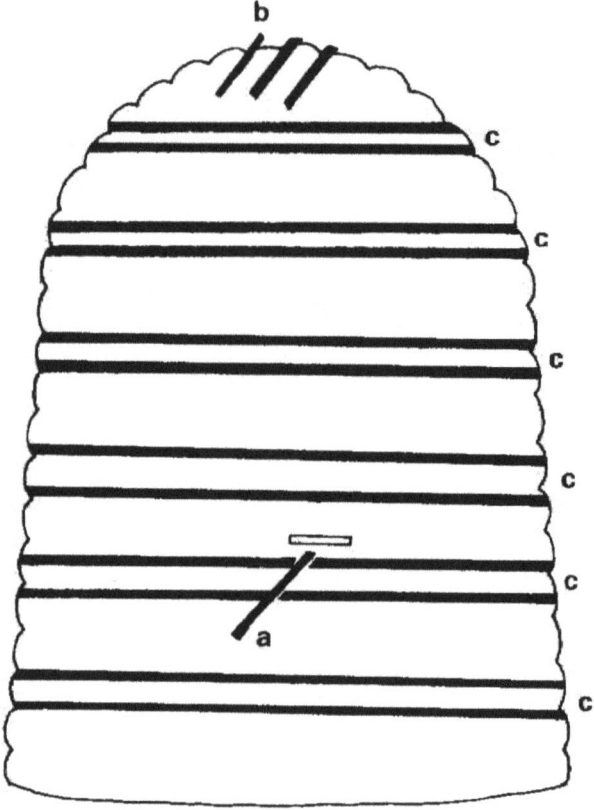

Fig 45. Skep to show position of bars a) bar below entrance hole b) top bars c) a series of paired, parallel, horizontal bars.

5. HANDLES

There are several ways to provide handles for baskets. We give a few examples and leave other possibilities to the ingenuity of the maker.

5.1 SIMPLE RAISED HANDLE

This is a simple handle located on the top of the bowl or basket, (see Fig 46) While we are weaving the last round of the basket, we first add a reinforcement. Then we lift the straw coil and wrap it several times. We can place these wraps so closely together than none of the straw can be seen. Then we bend the lifted part around and finally stitch it with a double stitch to the coil below (see Fig 47). To keep the opening that forms the handle when weaving we can put a piece of wood or a tube under it. We must also take into account that the other handle must be placed exactly opposite the first handle. We need to measure and mark this to make sure. Straw lends itself

Fig 46. Basket with a simple raised handle.

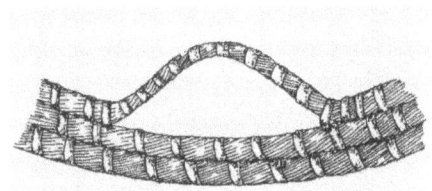

Fig 47. Simple raised handle.

well to a handle such as this, but purple moor grass needs to be crushed. We can reinforce the handle internally by inserting into its middle a piece of rattan cane or metal wire.

5.2 DIVIDED HANDLE

Taking into account what we have just discussed, it is easy to understand what we mean by a divided handle. When we want to put a handle on the top edge of a basket, we need to divide the bundle of straw.

The top part of the bundle becomes the handle and the bottom part becomes the extension of the last round. Straw must be added to both sections immediately after dividing so that they remain sufficiently thick, In addition, we must use a separate length of stitching material for the handle, which we have previously inserted into the straw bundle and secured (see Fig 48). Once we begin to rejoin the straw of the handle with that of the round, and begin to stitch, we find that the round is too thick. Thus, we will have to remove some of the added straw as we merge the bundles.

Fig 48. Divided handle.

5. HANDLES

We will also need to stitch in the extra stitching material which was used. This type of handle can also be attached to the outside of the basket (see Fig 49).

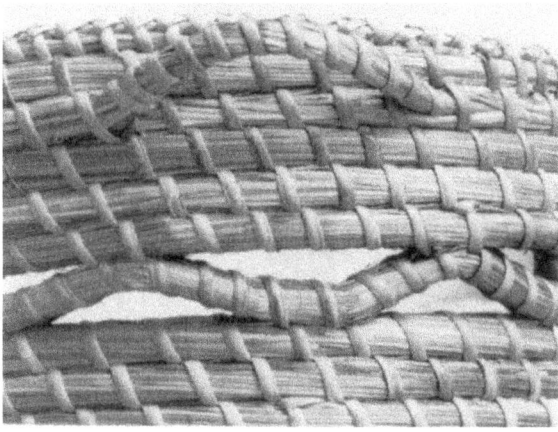

Fig 49. Divided handle on an open round and on the outside of the basket.

5.3 MOULDED HANDLE

A handle can also be fitted without using straw from the bundle. Before starting, we need some wire, or a piece of cane or willow, bent in the shape of a handle. Cane or willow will need to be soaked in hot water, before being bent into the required shape and held there by using string. After drying, it should remain in the desired shape. The moulded handle then needs to be stitched into the round (see Fig 50).

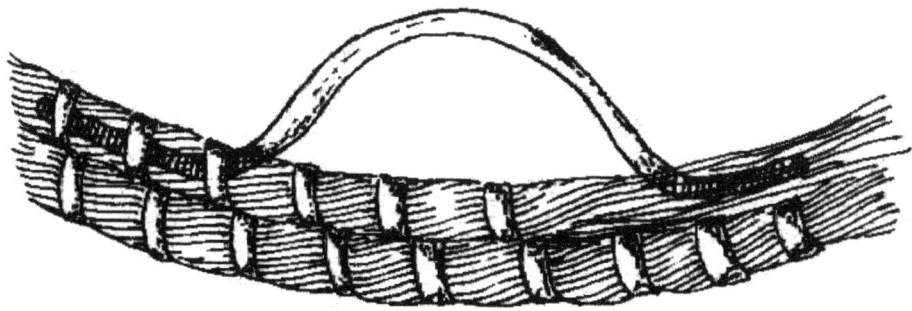

Fig 50. Moulded handle.

If the handle is too thin, we can wrap it round with stitches which are closely up against each other.

The handle in the top of the swarm-catching skep must also be pre-formed. The pieces to be stitched in should be bent in such a way that they follow the direction of the round of straw (see Fig 51).

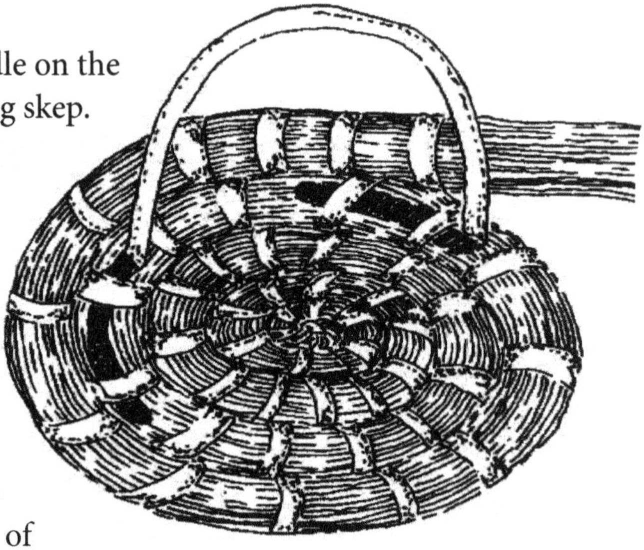

Fig 51. Handle on the swarm-catching skep.

This handle serves to hold the basket upside down under the swarm when catching it and then when the bees are inside, to transfer the skep. Instead of a sturdy handle like this, we could alternatively stitch a piece of rope into the top of the skep. The advantage of such a rope handle is that we can use it to hold the skep upside down to shake the bees out onto a sheet before hiving or to shake them directly into a hive. If we want to make a handle on the lid or the sides of a basket, we can do it in the same way, but here, a rope handle looks more attractive.

5.4. LID WITH A KNOB.

The lid is a separate braided disc which must have the same circumference as the top edge of the basket. It is fairly easy to fit a knob into the space in the centre of the lid (see Fig 52 and Fig 53).

5. HANDLES

Fig 52. Decorative baskets

Fig 53. Underside of lid (see Fig 52).

First, we make a knob from a round piece of wood and drill a hole through it from top to bottom. For appearance, the knob should be tapered. Before we attach it to the lid, we wrap the knob with the stitching material. This can be done most conveniently as follows. We insert a short end of the stitch at the top of the hole and pass the other end through several times. These stitches come to lie against each other and the short end of the stitch becomes automatically clamped in the inside of the knob. The hole in the knob needs to be reasonably wide, as the stitches take up quite a space. Once the knob is completely wrapped, we can cut off the remaining stitch and tuck any loose piece into the

5. HANDLES

hole at the base of the knob. Finally, we push the knob into the hole in the centre of the lid. A wooden pin (Fig 54), is hammered into the centre of the knob from below.

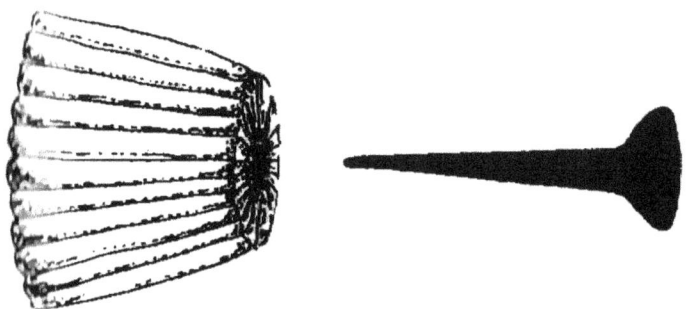

Fig 54. Wrapped knob and wooden pin.

You can easily make this wooden pin yourself. The end of the stitch that we pushed into the end of the wrapping stitches is now secured by this wooden pin. If you are anxious that the wooden pin will not be secure enough, you can put some glue in place before you hammer in the pin.

Speaking of the lid, we now need to give some pointers as to how to fit it. A lid will slide off the basket if it is just a disc. To avoid this, we need to weave a separate round at the base of the lid that fits neatly into the top edge of the basket (see Fig 53). A curved needle is most suitable for this (see Fig 13). We start this separate round with a thin bundle of straw, add in quickly and decrease towards the end, creating a round of equal thickness.

Another possibility is that we weave a separate double round on the outside of the basket in which the lid is recessed and thus rests on the former upper edge of the basket.

5.5 TIED-IN HANDLE.

A handle can't be made of straw or purple moor grass alone, as it would be too weak. It is quite laborious to tie a wooden handle, such as with old sowing baskets. For a wooden handle we should choose a flexible piece of wood such as willow or oak. A branch is flexible while it is still green. If we have a branch which would suit, we can, whether it's peeled or not, string it like a bow and leave it to dry. In a warm place, it dries in a short time and remains in the desired shape.

Before we tie the preformed handle, we drill a hole in each end, which will allow the braiding stitch to be attached. Alternatively, the ends of the handle can be thicker to allow a wide notch which will allow the stitch to be attached. We can tie the ends of the handle on the inside or the outside of the lid. Everything must be tight, or the handle may move or come off when we lift the lid.

The piece of stitching material that we use to attach the handle, should be tied in to secure it, either by using an S shape around a couple of stitches or by wrapping around a round of straw between two normal stitches (these alternatives are shown in Fig 55). Finally, if the handle is wrapped along its whole length with the braiding stitch, it looks very attractive (see Fig 56).

Fig 55. Two ways of attaching the securing stitches.

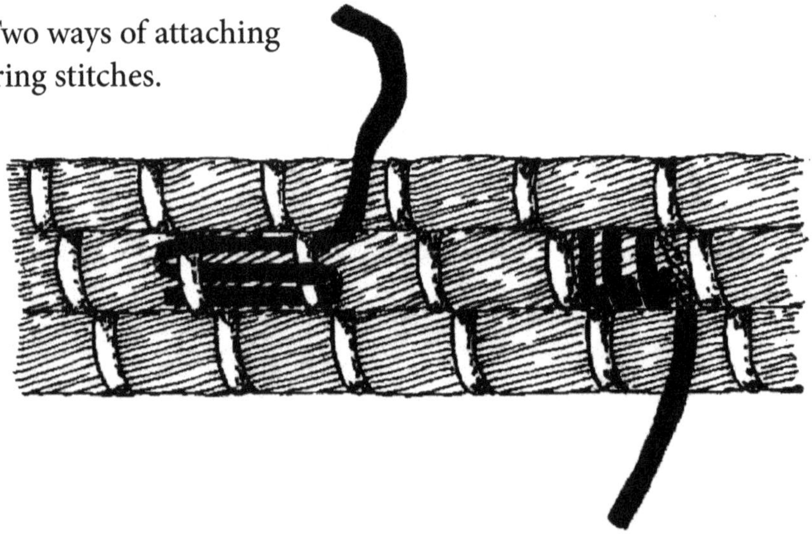

5. HANDLES

Fig 56. Basket with a braided handle and one with a handle made by dividing the bundle of straw.

5.6 BRAIDED HANDLE

A handle can be made in same way as previously discussed (see Fig 51 and Fig 56). As a core for the handle, we can use a thick piece of cane, a willow or oak twig or wire. If we use a twig, the ends that are stitched into the straw must be thin. Otherwise, an unsightly bulge will appear in the straw along the top edge.

We can cover the handle with straw or purple moor grass and then wrap this around with the stitching material, either closely wound together or spread out at the same distance as the other stitches.

5.7 HANDLE FORMED FROM THE ROUND

It is possible to form a handle by using the top round. This is an attractive form but rather laborious to produce (see Fig 56). For this we need two preformed and matching twigs or pieces of wire (see Fig 57).

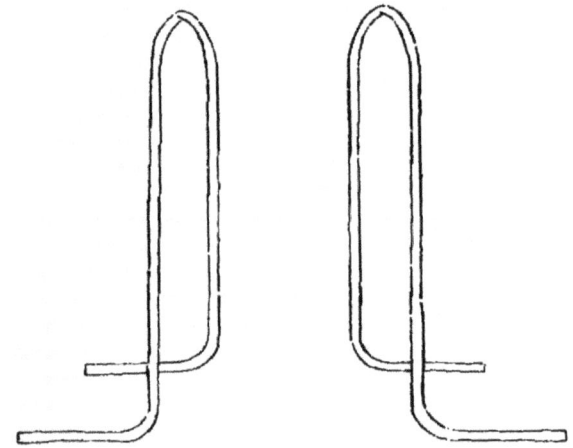

Fig 57. moulded handle

While working on the previous round we need to mark the two opposite places where the handle will be placed. It is necessary to measure or count the stitches if they are equally spaced. Now the hard work begins. When we reach the marking in the last round, we do not stitch the bundle further. We leave it with the remaining stitch for the time being (see Fig 58).

5. HANDLES

Fig 58. The top edge forms the handle

There needs to be a reasonable length of stitch available. If this is not the case, a fresh stitch should be inserted. Then we take a new bundle of straw, the same thickness as the round we have been working on. We also take a length of cane and insert a few stitches further in such a way that a long starting length remains. Starting with a quarter of the new straw bundle, we stitch it in and continue the last round (see Fig 58). The same must be done on the other side of the basket. The last bundle of straw should run naturally and be finished in the usual way. Only now can we fit the parts of the pre-formed handle into the twice broken straw round. First, we fix the handle on one side and then on the other, completely. This requires considerable manual dexterity. We take a (half) handle and insert one curved end under the last stitches of the straw round, pulling the remaining stitches tight and wrapping straw and handle (see Fig 59).

The stitching material must be secured at the end of this wrapped section or it will work loose. We do the same with the other half handle. Once one side has been dealt with, we move to the other side of the basket to deal with this part

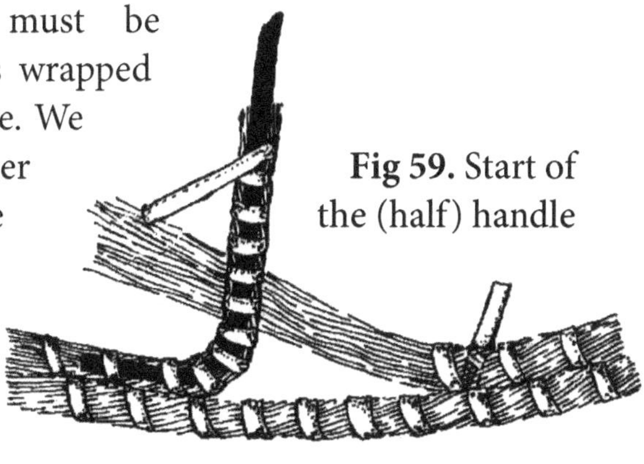

Fig 59. Start of the (half) handle

75

of the handle. After this laborious work, we start from the place where the handle should branch, winding the two half handles together. In this way all are brought together with the stitches, the preformed handle, the ends of the stitching material and the straw or purple moor grass. To get a nice-looking handle, we need to remove or add straw when necessary. Another possibility is to wrap and fasten each half handle separately from one side of the basket to the other. Then we have to join the two half handles with wrappings, so as to make one handle (see Fig 60). In this case, the wrappings need to be made so that they are tightly against each other, so as to hide the underlying wrappings of the two half handles.

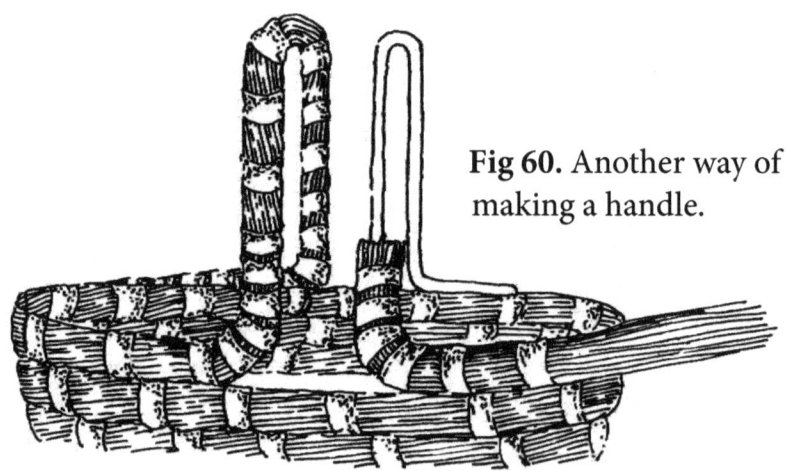

Fig 60. Another way of making a handle.

6. EMBELLISHMENTS

While spiral braiding is beautiful on its own, we can add a few more embellishments. First, a few special straw borders are discussed, then various decorations that we can make with the braiding stitches.

6.1 BOTTOM EDGE

Sometimes a separate bottom edge is woven under the basket. In addition to being decorative, this also serves as a reinforcement. (Fig 61).

Fig 61. Baskets (L with an extra bottom edge and R without the extra bottom edge)

Especially in the past, when baskets stood on stone floors and were frequently picked up, put down and moved, the stitches at the base were easily damaged by repeated rubbing. This is why a separate bottom edge was added, which could be renewed when it was worn through. Although modern baskets in our living rooms are not threatened in the same way, we should still describe this technique, because it is an important and beautiful element of the old basketwork (see Fig 6).

A bottom edge is stitched on under, or half against, the outer round of the base. When it comes under the base, we use a curved needle, because we have to stitch flat. We take a thin bundle of straw, flatten it and weave it into a few turns at the base. Then we quickly add straw and stitch the whole round. At the end of the round, we need to decrease quickly to finish what we started. We flattened the straw bundle at the start, so that the end and the beginning should look nice and symmetrical. This technique could also be used to place a double top edge in a basket to allow a lid to be fixed.

6.2 OPEN ROUND

An open round takes some effort, but the result is beautiful (see Figs 64 and 66). Going back to chapter 5, we can say that the round consists of a series of raised handles (See Fig 46) Before starting the open round, we need to count the number of stitches in the previous round. Based on this we divide that round into a number of parts that must be as equal as possible. The number of marked parts corresponds to the number of openings in the decorative round. If using purple moor grass, it must, because of the sharp bends involved, as we have said, be crushed and moistened. As soon as we lift the bundle, we make several wrappings without making any stitches into the round below. Then we bend the bundle in a curve and secure it to the round below with at least two stitches (see Fig 62).

6. EMBELLISHMENTS

Fig 62. Start of an open edge.

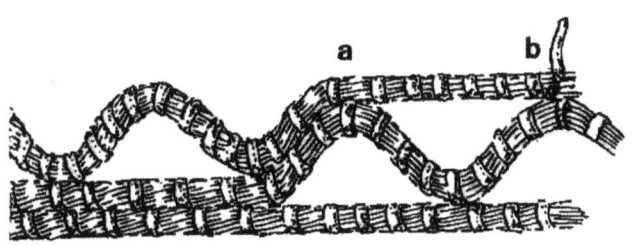

Fig 63. End of an open edge.

We can make the arches high or low according to taste. We can also reinforce the arches (and thus strengthen the whole basket) with wicker or metal wire. To ensure that the arches are the same height, we can use a round piece of wood or a tube placed in the arch and left there for the time being. We continue the decorative round in the same way until we get back to where we started. Here we stitch the straw bundle to the top of the first arch (see Fig 63 a). Then after some more wrappings we stitch the bundle to the top of the next arch and so on from top to top. (Fig 63 b) This decorative round can also be the last round so that it forms the top of the basket.

BASKET WEAVING

Fig 64. Knitting basket (with an open top, decorative features and divided handles).

Fig 65. Basket (open round finished with a quadruple stitch).

6.3 FOUR DECORATIVE EDGES TO FINISH OFF

We have chosen four reasonably straightforward possibilities from various options (see Fig 66).

Fig 66. Four Decorative edges to finish off a basket
a) a counter stitch, b) a triple counter stitch,
c) beekeeper's stitch d) a quadruple stitch.

Counter stitch

The counter stitch is a simple option for finishing off the top edge of the basket. First, we hide the stitch as much as possible in the top edge of the basket. We can tie a knot in one end of the stitching material and pull it into the straw bundle. Then we pass each new stich under the next old stitch in the opposite direction (see Fig 66 and Fig 67). If we insert the stitch so it is just below the surface of the straw round, we produce a graceful zig-zag line.

Triple counter stitch

If we have learned how to do the simple counter stitch, the triple counter stitch is not too difficult. We make three counter stitches overlying the original stitches, but each new stitch must miss out two of these underlying stitches (see Fig 68). With this decorative edge, we see the three counter stitches lying next to each other with a gap between the top stitches.

Beekeeper's stitch

The decorative edge that beekeepers often apply to the bottom edge of the hive is also intended as a reinforcement. The normal stitches of the lower edge of a skep are usually the first to be damaged and this causes the lower part of the skep to fall apart. Hence this additional protection. (see Fig 43). The beekeeper's stitch, like the previous one, consists of three stitches, running in the opposite direction to the original line of stitches. We also have to miss out two of the underlying stitches each time. However, the insertion is different [from the previous triple counter stitch]. The stitch must be inserted on the outside of the basket on the right under the winding and comes out on the left (see Fig 69). When we work like this, the three stitches appear as a graceful line on the top of the skep.

6. EMBELLISHMENTS

Fig 67. Simple counter stitch.

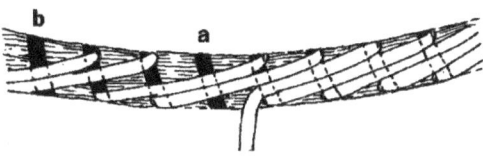

Fig 68. Triple counter stitch. In this, the stitch passes over the underlying stitch a, and then over the underlying stitch b.

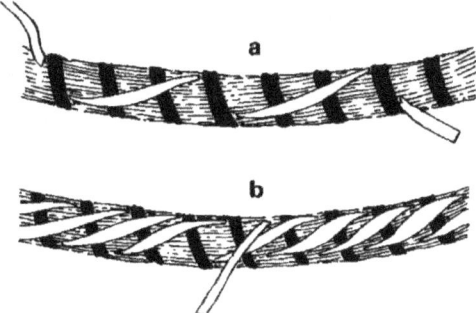

Fig 69. Beekeeper's stitch a) the first round b) the last stitch of the third round.

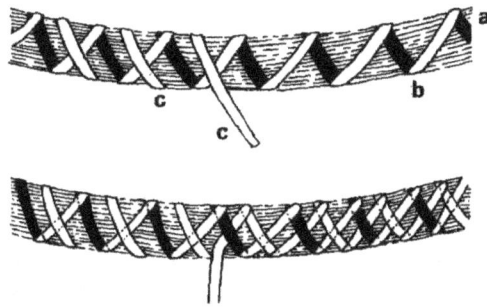

Fig 70. Quadruple counter stitch. Upper drawing shows a) original stitch b) first counter stitch c) additional wrapping. Lower drawing shows the second counter stitch, where the stitch is passed under the original stitch.

Quadruple counter stitch

This decorative border is named for the four stitches that are braided together. It only uses stitches that we are familiar with. We need three stitches one after another, not all at the same time. First, we make a simple counter stitch. With the second stitch, we simply make a new wrapping around the whole straw bundle (see Fig 70 upper).

With the third stitch, we make another counter stitch, but this time the stitch goes under the added stitches (see Fig 70 lower and Fig 65 and Fig 66).

It is also possible, starting from the four examples described, to devise and implement more complicated and even more beautiful decorative edges. Finally, it should be noted that all the decorative edges that we've described can also be applied to the side of the separate bottom edge or to the outer edge of a lid.

BIBLIOGRAPHY

Den Naerstigen Byen-Houder, onderrechtende hoe men met nut en profijt de Byen regeeren en onderhouden sal. (The diligent bee-keeper, enlightening how one should profitably govern and maintain bees). Amsterdam, 1686

JJ Speelziek, *Korfimkerij (Skep beekeeping)* Teuge, 1995

E Jacobs and HWM Plettenburg, *Der oude imkerij, 3e druk, uitgave van het Rijksmuseum voor Volkskunde, (Old Beekeeping, 3rd edition, published by the National Museum of Folklore),* Het Nederlands Openluchtsmuseum, (The Dutch Open-Air Museum), Arnhem 1978

H van de Kammen *Bijenteelt met korven (Beekeeping with skeps)* in Bijenteelt, Maandblad van de Bijenhoudersbonden (Monthly Magazine of the Beekeepers' Unions), (56) 1978, 68-70

J Weyns *Spiraalvlechtwerk uit de Kempen* in Nederlandsch Tijdschrift voor Volskunde (*Spiral weaving from the Kempen*) in the Dutch Journal of Folklore 39 (1934-1935) 137-148

J Weyns *Spiraalvlechtwerk uit de Kempen, aanvulling*, in Nederlandsch Tijdschrift voor Volkskunde (*Spiral weaving from the Kempen, an update*) in the Dutch Journal of Folklore) 51 (1950) 118-126

COLOPHON

WITH THANKS TO

The graphical staff for the first edition of this book, published by Cantecleer, de Bilt, 1979, from which photographs and drawings have been copied:

Hans van Ommeren (photographs) and Mary Nauta (drawings)

Janus Damen of Dongen (Fig 40, the swarm catching skep)

National Museum of Folklore (Dutch Open-Air Museum) in Arnhem (Fig 6 and Fig 7)

PPO Bee Sector, Ambrosiushoeve in Hilvarenbeek (Fig 39 and Fig 44)

Roel ten Klei, of VBBN, who took the initiative to republish this book

ADDITIONAL IMAGE CREDITS

Fig 3. The Beekeepers, Peter Bruegel the Elder 1565. Wikimedia Public Domain. *https://commons.wikimedia.org/wiki/File:Pieter_Bruegel_the_Elder_-_The_Beekeepers_and_the_Birdnester_-_WGA03528.jpg*

Fig 4. The Lovesick Maiden, Jan Steen. Wikimedia Public Domain. *https://en.wikipedia.org/wiki/The_Lovesick_Maiden#/media/File:The_Lovesick_Maiden_MET_DP147594.jpg*

Fig 5 Beaker Netherlands Wikimedia Creative Commons 3.0. *https://en.wikipedia.org/wiki/Bell_Beaker_culture#/media/File:Beaker_Netherlands_1.jpg*

RESOURCES

(AS OF FEBRUARY 2025)

Rattan cane (The most suitable is 5mm glossy lapping cane).

Seat Weaving Supplies, West Moors, Ferndown, Dorset. Tel 01202 874737 https://www.seatweavingsupplies.co.uk

Somerset Willow Growers, Bussex Farm, 65 Liney Road, Westonzoyland Nr Bridgwater, Somerset, TA7 0EU https://www.willowgrowers.co.uk

Tools

A useful tool for threading cane or other binding material through the bundle of straw is a 168 mm Swedish Splicing Fid Marine Super Store https://www.marinesuperstore.com

Sourcing straw

https://www.strawcraftsmen.co.uk/suppliers gives details of suppliers of un-baled long wheat straw, but probably the best approach is to find a farmer who grows long wheat, (Maris Widgeon or an older traditional variety), and ask if you can cut a small amount just before harvesting. The straw is best cut by hand with shears. Before drying and storing, it will need the leaves removed and the grain taken off (to prevent mouse damage). If kept upright in cool and dry conditions it can last for years before being used.

Books

Frank Alston, (1987), *Skeps Their History, Making and Use*, Northern Bee Books, ISBN 0- 907908-38-1

IBRA, (2009), *Skeps. Tools and Accessories*, The IBRA Museum Part 2, ISBN 0-86098-360-2

Arthur Stanicroft, (2008), *Straw and Straw Craftsmen*, Shire Library, ISBN 078-0-7478-0103-0

BASKET WEAVING

RESOURCES

BASKET WEAVING

www.ingramcontent.com/pod-product-compliance
Lightning Source LLC
Chambersburg PA
CBHW040318170426
43197CB00021B/2956